To/. Helen,
Best of luck
in your career,

Maintenance & Facilities Management

Regards,
Brendan
17/05/16

Brendan Shine

bjshine@avara.com

ORIGINAL WRITIN

978-1-908282-51-4

A CIP catalogue for this book is available from the National Library.

Published by Original Writing Ltd., Dublin, 2011.

Printed by Cahill Printers Limited, Dublin.

This book is dedicated to my daughter Sophie

Acknowledgments

My thanks and gratitude goes to the people and organisations that encouraged me to write this book, especially for their valued suggestions, personal insights, shared experiences and help.

Foreword

The purpose of this book is to provide the reader with a fundamental understanding of all the various aspects involved (using soft and hard skills) in becoming a good engineering practitioner in an industrial, aerospace, marine, commercial, agricultural or domestic setting. The following chapters cover the tribal institutional knowledge (i.e. passing down of information from master to apprentice) that must be maintained across all engineering disciplines.

Engineering savvy is the collection of knowledge about one's industry and line of business in general and how all the different areas smoothly interact with each other in making a good quality product i.e. Electrical, Mechanical, Chemical, Automation technology, Programming, Electro-Technical and Physics etc. This type of critical knowledge is sometimes difficult to document and sometimes cannot be imparted by an instructor or a manual. It is crucial personnel familiarise themselves with the company's operations beyond their own departments and use the power of knowledge to help them succeed.

Managing and retaining such experiential knowledge is becoming one of the major challenges facing companies in the near future. Experienced engineering personnel, who are resident in any workplace over many years and maybe, are about to leave or retire from the business can sometimes predict or diagnose a problem that even the vendor of the equipment or the designers of the entire installation did not envisage.

An experienced person, who possibly has spent their whole career with a company, may have accumulated vast knowledge and experience about complex systems and processes. Keep in mind, that this 'know how' will just disappear when they retire. Companies must address this challenge early on and ensure this valuable and irreplaceable experience is not lost.

The knowledge vacuum must be filled. It's vital that existing dedicated onsite personnel learn or come up to speed with the experienced person's level of knowledge, otherwise, it will present itself in costly process interruptions which can put equipment and others at risk. A company should consider a part time retainer for the person who is retiring to ensure plant operations continuity while others take over the role. Some companies are having trouble attracting engineers with the right level of competence for these specialist roles.

Sometimes complex and seemingly fail proof technical systems can go wrong because of overlooked problems that interact with each other in unexpected ways.

Automation technology will provide the relevant technical information needed to monitor and operate equipment safely. It exists not to replace people, but to assist them, it cannot entirely take into account 'the human factor'. A dedicated control system will send 'warnings' (can be an audible alarm, a visible flashing light, written text on a PC screen or all together) of impending danger but cannot make a person perform an action to counteract this danger.

The control system can only do what it is programmed to do (alarm, put the equipment into 'hold', 'fail safe' or 'shutdown mode'). If any action occurs outside what it's designed to monitor and control, the 'human factor' has to play its part. Personnel must intervene and take swift action, which is the very reason why a person's specific relevant knowledge, experience, skills and abilities are vital to any particular dedicated control system.

N.B. Every effort has been made to ensure what's written in this book is accurate. Personnel must always ensure when carrying out operations and maintenance of equipment that the regulations and guidance issued by their country's authorities are consulted and implemented at all times.

The following chapters are an accumulation of many years of personal 'on the job' engineering experiential knowledge and is a synopsis of the expert training gained, off site college, HR training courses & seminars attended. It has been written to assist Managers, Engineers, Tradesmen, Apprentice tradesmen or the D.I.Y. specialists.

About The Author

Brendan Shine has over 30 years experience (including electrical, electronic, pneumatic and hydraulic qualifications) in the engineering industry. He qualified as an industrial electrician in 1984, from there moved into a maintenance and operational role for many years which involved maintenance and operations of all on site equipment, then in 1992 moved into the calibration and industrial automation side. Returned to college in 1998 and qualified as an electrical engineer and moved into a project engineering role in a technical services division. He is currently employed as the Maintenance and Facilities Manager of a very large and successful pharmaceutical company who have been in business for over 35 years.

Contents

1 INTRODUCTION

In today's industries, tight budget control, cost cutting implementation, increasing efficiency and freeing up capital is paramount to a business's survival. Replacing a piece of expensive equipment because of its age is no longer an option.
Companies are taking stock of the equipment they have on site, how they can properly maintain and modernise versus replacing and still make the equipment as efficient as possible, thus reducing overhead costs.
Equipment designers are trying to remove the 'black art of fault finding' by using technology to do self diagnostics and flag potential problems on the machine itself. Automation is constantly evolving to assist non technical personnel in diagnosing faults while at the same time reducing a company's dependence on a small group of specifically trained personnel to keep the plant operational should they be on holidays, on sick leave or actually leave the company.
Other innovation, such as 'Start up Assistant' is designed to aid commissioning. 'Diagnostic and Maintenance Assistants' are now being provided by machine manufacturers via 'Intelligent Controllers'. State – based control and self learning diagnostic routines, have raised a machine's controller ability to detect, annunciate and describe problems with machinery. Take the automotive industry for instance and how far it has advanced in 40 years. From basic analogue speed dials, oil level indicators, temperature gauges to EMS's (Engine Management Systems) with full digitally operated monitoring and detection safety systems which assist the driver in the smooth, efficient and safe running functions of the vehicle itself e.g. ABS braking, parking detection systems, optimum fuel efficiency hybrid cars, cruise control etc.

Another key aspect to reducing costs and downtime is to train in-house personnel to do 'specialist' work. The initial cost of the training may be high and possibly will involve sending people off site for a number of days to receive it, but, it will be well worth it in the long run.

When personnel are responsible for maintaining a piece of equipment, they must take over 'due care', 'custody' and 'control' of all its systems and take ownership, be accountable and responsible for all activities associated with the piece of equipment during its lifecycle.

Trouble shooting and fault finding methods are disciplines that must be learned, managed, practiced and adhered to. If adopted and implemented properly, will save many lives, prevent environmental & industrial accidents, save industry massive monetary losses per year in personnel injury claims, machinery down time and lost productivity. Knowledge and experience are vital elements of any trouble shooting & fault finding methods. **Learn from others**, their knowledge and experience will prove invaluable. Learn from their successes and implement them. Just as important, learn from their mistakes and try not to repeat them. It is inevitable personnel will make mistakes, but, they have to be aware, admit and learn from them and try to ensure they are not repeated.

> *A person must think before they do, to question is to understand.*

Good literacy and numeracy skills are essential to any engineering discipline. One of the single most important actions any person carries out as part of their daily work routine is signing their name and date to any specific work completed. This is where a considerable amount of errors emanate from and subsequently becomes a 'knock on' effect for other parameters being recorded i.e. writing is not legible, wrong date and time is written in etc. The key is to stop, think and perform a quick scan of the document before it is signed and dated. **If that type of thinking can be propagated to all day to day work activities,** deviations and mistakes will be immediately reduced across any business.

Personnel may be faced with a situation that needs resolving as soon as possible, there may be only anecdotal information available. Once they have good knowledge of control systems and equipment in general, their natural trouble shooting and fault finding methods should take over and be apt to try new approaches and seek creative new solutions to a problem. They must learn to adapt to whatever situation arises, keep an open, ever learning mind where new thinking may be required and not become entrenched in the way they always did things. **New methods** of trouble shooting and fault finding techniques will need to be adopted as technology and equipment is constantly advancing. **The fundamentals, though, will never change.**

> ***Note: Don't take risks, if in doubt "ASK". Stop and seek advice.*** **Don't be afraid to ask dumb questions, they are easier to solve than dumb mistakes.**

The following chapters are guidelines and may contain terms or references the reader may not have heard of, but can be accessed via the internet or a local library. There is an abundance of sites where if needed, further information and research is readily available. Where terminology or abbreviations are used, such terminology is listed in the 'Glossary' section at the back of this book and an explanation provided as to its meaning.

Tip: When searching the internet for relevant pieces of information in any one of the search engines, to narrow a search and find the information faster, type in the words 'Diagrams/Images of' in the search box icon.

E.g. **Diagrams of open and closed loop control** – an excellent selection of diagrams and other links associated with the subject required will become available.

2

People Skills

Having technical skills and knowledge is a prerequisite for any engineering related activity. Equally as important though, is that personnel acquire the relevant people skills needed to help them interact professionally, efficiently and effectively with people within their own department and with other departments within the company. The same applying, when it comes to outside vendors and clients, in the pursuit of maintaining the company's goals and objectives. There are 3 areas, which, if addressed properly, will help any business to be more successful, they are:

1. **Communication**
2. **Open mindedness**
3. **Team work**

1. Communication is the 'creation of understanding' by the transfer of knowledge using verbal, past experiences, stories, script, illustrations, examples and will play a key role in any engineering discipline.

When a relevant message is being verbally conveyed either in a 1:1 situation or in a team brief, try to stop talking and listen, one cannot multi-task speaking and listening. When a person is thinking intently about what they want to say, they're actually not listening to what is being said. Don't wait for information to be supplied, go after it. How specific knowledge is accumulated, identifying the people who have this knowledge and how to go about attaining it from them will determine how easy or hard a task is going to be. The first challenge to be faced is how to use this specialist knowledge in order to complete a required task and secondly, maintaining effective communication with the person who assigned the task in the first place.

2. To ensure a good working environment, personnel throughout the company should try to be **'Open minded'**, as it's important to realise that few people are the way they would like them to be. Showing diplomacy and being sensitive to other people's points of view can make solving a problem much easier to resolve. Never use aggressive or confrontational methods to get information. Ask for advice, support or guidance when in need. Don't assume others will notice when this is the case. Work colleagues will be focussed on their own tasks and maybe haven't a lot of spare time to be answering questions. Sometimes, although rarely, a person may have to deal with an aggressive, short tempered person who may dismiss them when they ask for information. **Remember,** stay calm but at the same time responsive. Have the right mindset of resolving the issue at hand. When faced with a difficult person, it is easy to want only to be right. Acknowledge there may be a need to compromise to progress forward. **Pick the right moment,** become aware of how other people react when approached i.e. they could be having a bad day and are irritable. All the required information may not be on site at all times, knowing where, who to contact and how to source it as quickly as possible is key.

Tip: Other than safety or emergency situations that may need immediate attention, **don't do work that wasn't assigned to you by your direct boss**. Adhere to the 'Chain of Command'. If you have 3 or 4 people telling you what to do and only answer to 1 of them (it may be the General Manager, another department manager or supervisor) tell them that you will gladly carry out their requests. But first, ask them to channel 'what they want done' through your direct boss and he/she will prioritise the work accordingly. That way you keep everyone happy and your work life will be a lot easier.

Differences in perspectives and priorities can lead to differing views on the contributions a person makes at work. If given a list of work to be completed within a certain time frame and difficulties are been experienced completing the tasks because of factors outside a person's control, keep the affected manager informed of the situation. Ask them for direction, don't assume that they already know problems are being encountered.

3. The benefits of **'Team work'** cannot be under estimated. A person cannot do everything themselves. Accepting help or delegating responsibility is not a sign of weakness, but a growing process that helps progress people both personally and professionally. Personnel may only be just starting out on their career or are already at the upper-management level. It doesn't matter how smart or capable they are, **if they work with others,** they will find that the overall results come much faster.

Whether in a 1:1 meeting or a group session with other team members, a person must have the ability to communicate clearly, specifically and unambiguously no matter how frustrated they may become at other people's apathy or lack of knowledge. **Never** be disrespectful, inconsiderate or 'talk down' to others. **Try to communicate in person,** emails are an excellent means of communication and transferring files, information etc, but, it's always better to interact with people 'face to face' where possible.

Reserve email for those not in the general workplace. Pick up the phone instead or leave a place of work to deliver the message. It is vital to get along with others, sometimes personalities clash, but, stay professional. It leads to a better working environment, when problem solving and making better judgement calls.

Tip: Have you ever said something or sent a message in anger and regretted it afterwards? Before making a possible bold contentious statement in a meeting or in a 1:1 discussion with members of your team, peers or manager, consider the possible repercussions (if any) and can you handle them? If your instincts tell you not to say it, trust them. Discuss it with a trusted colleague first and ask them for their advice and try to establish do 'they see it, the way you see it'. The same applies when sending a controversial email; get a trusted colleague to 'proof read' it first before sending. If you are left with no alternative and have to say or send something that may cause offence or a negative reaction, try to adopt the 'Sandwich approach' technique: **"Give a person a compliment, make your comment, then give another compliment".** The message is delivered between 'two pieces of niceness'.

Proper people skills techniques must be honed, practiced and implemented. Personnel across the company should be encouraged to read as many articles as possible on the subject, especially 'Emotional Intelligence', it will be time well spent.

Think positive – Having the **Right 'Can do/Will do' Attitude'** helps a person to think, solve problems, make decisions and be creative. It can help in making an objective appraisal of any situation and less likely to make mistakes. Avoid making the same mistake twice, personnel must ask, can they avoid or go around the problem and try to turn it to their advantage. The ability to make effective decisions about what to change and how to change it during a crisis separates a person who has done their research, armed with theory, knowledge and experience from the rest. Have confidence in yourself and you'll inspire others to have confidence in you.

Learn to **manage** a given situation. Plan for the unexpected, this will keep surprises to a minimum. If you've thought of the things that could go wrong with any proposed work planned, confident decisions on corrective actions can then be made when necessary.

E.g. You are a team leader of a crew of 10 personnel, all involved in resolving a major problem. Regardless of how large or small a problem maybe, it's not the problem that matters; it's how you respond to it. Stay focused, show leadership, confidence, control, patience, organisation, strategic planning and be able to shift gears to quickly find a solution to a problem regarding personnel and equipment.

Give proper clear direction and work through each issue until it is resolved. Be aware of any potential dangers, maintain a 'heightened situational awareness' at all times. Whatever environment you or your team are in, be ever mindful of changing conditions i.e. breathable air conditions, deteriorating weather conditions, tidal conditions or heavy machinery moving into the area. Never put the lives of your team or your own at risk regardless of the time constraints or rising costs involved in trying to get the problem resolved.

Don't blame others for your mistakes, own up to any errors you make. Offer explanations, not excuses. Try to fix problems and don't let pride keep you from speaking up and seeking advice from others if you can't resolve an issue. Take responsibility for your decisions and actions, you'll be respected for it, and maybe even set an example. The team must have confidence in you and your abilities to lead them in normal day to day activities or through a crisis. If they think you are reckless and take unnecessary risks regarding their welfare, safety or the work to be carried out, they may not follow you.

Remember, actions speak louder than words, always ***'lead from the front and protect those who are behind you'***. It's better that the team follow in your footsteps willingly, rather than having to drag them along, begrudgingly. Accept that it is inevitable that one of your crew may make a mistake, if they are to be reprimanded, do it in a 'one to one' scenario, never in front of a group of people, you will lose their respect and that of those around you. Tact is the art of making a point without making an enemy.

Don't be intimidating towards the team, you'll find you will never get a straight answer and end up defeating your own purpose. Nothing lowers the tone of a conversation more than a raised voice, a good argument has no need of a loud voice. In saying that, don't ever be afraid to challenge them either. Let them know the standards that need to be maintained at all times regarding:

- Being responsible for their actions.
- Health & Safety Practices.
- House keeping practices.
- Personal Attitude.
- Professional behaviour.
- Professional capabilities of their actual 'day to day' work activities.
- Interaction and communication with other team members.
- Willingness to share 'work related' information.
- Being helpful to others.

Being a team leader needs all of the above attributes, but, just as important is having knowledge on disciplinary and grievance procedures.

Try to ensure to have the right mix of people on the team, who maybe are stronger in certain areas than you are. Let them get on with the job and then act only in an advisory or overseeing role.

Learn from mistakes and get acquainted with as many talented people as possible. The key is to learn from others' experiences.

3

Setting up a Maintenance Department Structure

An Engineering department must be committed to maintaining the equipment and site facilities which sustain a company's business needs. This must be done in a safe, regulatory compliant manner by determining the best, most cost effective, competitive and sustainable way.

The engineering department must be committed to the objective of employing the best available practices to ensure all team members are competently trained on all aspects of their respective jobs.

Engineering asset management of ageing equipment requires effective maintenance investment management. Precision maintenance is synonymous with asset reliability. New, more creative approaches are needed to make budgets go further and minimise costs and waste.

Improving reliability of equipment can improve profitability by preventing costly shutdowns i.e. eliminate costly business interruptions. An essential part of any business plan must be to provide new solutions in reducing energy consumption, minimise downtime and to react intelligently to real problems as they occur.

A Maintenance Practitioners' key objectives are to:

- Ensure proper equipment maintenance is carried out at all times as inadequate or inferior maintenance practices often cause a slow, irreversible equipment degradation.

- Prevent equipment failures.

- Carry out the 'day to day' tasks and deal with the inevitable crises and issues that will arise.

- Maximise the uptime of the business's processes by ensuring quality reliable equipment is installed with redundancy thus ensuring optimum lifetime of all machinery in a predictable, safe and low cost manner.

Good Management (Soft Skills):
There are many facets to a good management structure and to the future success of a business. Ensure the 'Engineering Manager' in place:

- Is suitably qualified, knowledgeable and experienced in the key engineering disciplines i.e. Electrical, Mechanical, HVAC, Refrigeration systems etc.
- Knows how the business operates as a whole and understands what other departments do. This will put him/her in a better position to style their work to be of maximum benefit to the business itself.
- Has good personal people skills and is a good communicator.
- Has good organisational skills and can handle the inevitable pressure and stress that can accompany such a role.
- Has good conflict management skills.
- Instils a good, honest work ethic and is a good decision maker, 'they need to say what they mean and do what they say', to do that; they need to practice what they preach.
- Is calm and responsive, addresses others with respect and can talk ***to people, not at them.***
- Has the ability to motivate their staff and, in turn, get them to maximise their (and the business's) productivity in tandem with using good 'Time Management' techniques.
- Asks team members what obstacles (if any) block their ability to commit or participate fully in their 'day to day'

activities and work to remove the barriers and address the issues.

- Is capable and not afraid to make unpopular decisions and taking the necessary steps to implement them to achieve the company's KPI targets.
- Respects everyone's opinions, this will greatly improve their odds of having a successful, positive working environment.
- **E.g.** Nobody wants an autocratic boss hovering over them, micromanaging and watching their every move. It is up to the manager to praise employees' good work and offer suggestions in a professional, civilised manner.
- Employees want to emulate the 'higher-ups', so set the standard of expectation. Even if you're the one in charge, others may not automatically see you as such. In order to get others to accept you as their leader, you'll need to prove that you're worthy of your status. People in general, admire and appreciate another person's achievements, not the 'title' they have. As a manager, you may want things done your way, but you can only guide others in the direction you want them to go. When you treat staff like servants, you eat away at their pride and self-esteem, and they will ultimately despise working under you.
- When you entrust a responsible and capable person with an important task, give them some breathing room to perform it. Asking team members to come up with a new project means allowing them to get their creative juices flowing without constantly barraging them with questions or insisting they do things your way. Step back and demonstrate that you have confidence in them, this will allow for a more efficient and less hostile working environment.

E.g. You may have multiple team leaders working on a range of projects at any one time. Some team leaders may like the title of 'team leader' but maybe are unsure about their 'decision making and people skills'. As their manager you must reassure them how certain you are of their capabilities and you are available to give direction and advice if need be, at any time. Value your teams opinions, being a leader leaves much room for interpretation, sometimes, what ***you*** think is the best answer to a given dilemma might not necessarily be the best way to proceed according to those around you.

As a result, accepting a leadership position includes having to make difficult choices, and being able to accept that someone else may come up with a better plan of action. A great leader makes choices based on the business's best interests, not their own.

Just because you're the one in charge, doesn't mean you know everything about the company. While you may be more experienced and can likely provide more insight than others, you should always strive to learn new things.

Performance Management is vital to any organisation where:

1. Work is planned and expectations are set.
2. Performance of work is monitored.
3. Staff ability to perform is developed and enhanced.
4. Performance is rated or measured and the ratings summarised.
5. Top performance is rewarded.

Performance Management involving 1:1 meetings must be held at least once every six months between the manager and all individual team members (monthly meetings with new employees for the first 6 months and beyond if necessary). Direct feed back can then be given on their past 6 month's performance using these five key discussion points as reference. The time dedicated to this task cannot be underestimated and is real 'value added'. A dedicated folder or soft copy should be created, maintained and updated by the supervisor/manager as required for each team member which will include the normal 'day to day' department activities. The following points can also be used as peripheral discussion points which may need to be addressed:

- Absenteeism.
- HR issues.
- Time keeping.
- Department Accidents/Incidents.
- Department Compliance issues.
- Good Documentation Practices.
- Training attendance.

A team of well trained professionals is the key to running a good engineering department. The manager must make sure all their needs are catered for as much as possible both personally and professionally. Understanding the importance of maintaining good team morale, trying to ensure their needs are met and the implementation of a good team work environment cannot be stressed enough. Whether it's by:

- Salaries and remuneration packages.
- Maintaining modern clean toilet/shower areas /canteen/ workshop area.
- Providing proper PPE.
- Suitable 'fit for purpose' water proof clothing.
- Purchasing upgraded quality tools and equipment.
- Increased training.
- Applying new technology.

A manager must show that they are willing to fight for their team. They won't always be successful, but it's important that he/she acts as their advocate.

Good Engineering Management (Hard Skills):
The following fundamental structures should be in place:

- **A** **Project Management and Original Equipment Manufacturers.**
- **B** **A properly indexed and 'User friendly' Engineering library.**
- **C** **Good Budget Strategy.**
- **D** **Good Service Level Agreements with vendors and contractors.**
- **E** **A well stocked and efficient Engineering stores area.**
- **F** **Vendor 'Maintenance Software' System Package.**
- **G** **Vendor 'Calibration Management Software' System Package.**
- **H** **Holistic site Energy Management Strategy.**
- **I** **Emergency back up systems and contingency plans to ensure business continuity.**

A. Project Management.

An engineering **'Turn-Key Project'** is a project that is scoped, designed, built, installed, commissioned and handed over to a company in a fit for purpose, versatile, reliable, energy efficient and ready-to-use condition. In the following example, an expertise is required which is not available from a single individual; a suitable turnkey project provider is now selected and employed by the company.

E.g. a new diesel powered 1,000 kVA electricity generator is to be installed in a plant to supply 'back up' essential power to critical onsite equipment in the event of mains electricity power loss from the incoming national grid system. The various works will involve structural, mechanical, electrical and IT expertise to install.

A turnkey project such as this involves the following elements depending on scope and size:

- Project administration
- Design and engineering services
- Subcontracting of works
- Procurement and expediting of materials and equipment
- Inspection of equipment prior to delivery
- Management of actual equipment installation
- Control of schedule and quality
- Commissioning and completion
- Performance-guarantee testing
- Handover Documentation Package.

The turnkey provider provides the following warranties and guarantees and accepts liabilities which include:

- Warranties for the timeliness of deliveries of equipment, of erection and of completion times of structural, mechanical and electrical works.
- Warranties for workmanship according to specifications and guarantees that correct standards will be used.
- Liability for property and equipment.
- Correct safety standards being implemented.
- Contractor undertakes to assure that Mechanical and Electrical performance will be maintained for a definite period.
- Warranties and guarantees that the complete installation is suitable for purpose and fit for use.

Project Management (PM) companies and **Original Equipment Manufacturers** (OEM's) welcome early involvement in the design process, allowing a company to get advice from their knowledgeable and experienced experts **who:**

- Provide a family of services delivered by a multi-disciplinary team of Design, Structural, Electrical, Mechanical, IT and control professionals.

- Understand the full capabilities of modern control systems and how to ensure a company achieves the optimum cost/performance balance of installed equipment.
- Act as a focal point for the project, coordinating information and technical capabilities.
- Can assess, design, implement, audit and manage new and existing control and information networks and the security technology, policies and procedures for those networks and the personnel that use them.

Original Equipment Manufacturers must help maintenance personnel by ensuring their machines and associated systems are easy to troubleshoot, modify and repair during the machines lifecycle, in other words, 'maintainable'.

B. Create an Engineering Library

A technical file should be compiled for all projects completed by the Engineering Projects Department. Technical files are created to provide a historical archive for future reference. A copy is issued to the System Owner on completion of a project, with a master copy retained in the site library archive room.

Compiling a Project Technical File:
A typical contents page for a Project Technical File describes the various documentation requirements included under the relevant headings.

Approved Documentation:
Includes signed copies of the following:

- Project Initiation Form (PIF)
- Scope Documents (e.g. URS, FDS etc.)
- Copy of approved Change Control (ref. 'Site Change Control Procedure')
- Copy of approved Capital Request Forms
- Any other miscellaneous approved documentation

Drawings:

- System "As-Built" P&IDs
- System "As-Built" G.A's (General Arrangement Drawings)
- System "As-Built" Electrical and Instrumentation Drawings / schematics
- System "As-Built" Mechanical Drawings
- Other Miscellaneous Drawings relevant to the project

Commissioning Test Packs for:

Mechanical Systems:

- Commissioning Reports
- As-Built ISO's
- Pressure Test Certificates
- NDT (Non-Destructive Testing) Certificates
- Welder Competency Certificates
- Material certificates
- Miscellaneous

Electrical & Instrumentation Systems:

- As-Built Electrical and Instrumentation Schematics
- As-Built Instrument Loop Sheets
- Electrical Bonding and Electrical Test Sheets
- Equipment Data Sheets

Automation Systems:

- Control URS / FDS (As-Built)
- System Architecture Schematics
- (SDS) Software Design Specification
- Commissioning Test Protocols
- FAT / SAT Reports

Certificates of Conformity:

- Calibration Certificates
- Compliance Certificates
- PED (Pressure Equipment Directive) Certificates

- Material Certificates of Conformity
- Miscellaneous Certification (e.g. Noise)

Purchasing Data:

- Project Purchase Orders
- Purchasing / Vendor Data
- Equipment and Instrumentation Data Sheets
- Miscellaneous

Maintenance:

- Operation & Maintenance Manuals
- Calibration List / Updated Master Instrument List
- Equipment Index
- Spare Parts Lists
- Warranty Certificates
- Miscellaneous

EH&S Data:

- HAZOP / Risk Assessment Data
- Miscellaneous

Correspondence / Minutes:

- Project Handover Record
- Miscellaneous
- Labelling of Project Technical File
- Project Technical File Distribution

As part of the project handover an Engineering Project Handover Form must be completed and included in the Project Technical File.

Hardcopies of the Project Technical File should be made and circulated as required to:-

1. The System Owner
2. Engineering / Maintenance
3. Library Archive (One Copy Retained)

A soft copy of the Project Technical File is maintained on the Project Engineering Hard Drive.

Compiling a Maintenance O&M technical file:
A typical contents page for a Maintenance Technical File describes the various documentation requirements included under the relevant headings. Normally this file is kept in the maintenance office/workshop area.
When creating a maintenance hard copy folder of relevant associated information pertaining to any piece of equipment on site for other personnel to reference or use, the following format can be used using typical chapter headings and then subsequently stored in the engineering library for easy retrieval. All information, if possible, should be electronically scanned and stored in an easily accessible 'soft copy' folder on a company's Maintenance Engineering shared 'hard drive' for easy access from any site PC as needed. This may seem very time consuming and too costly to carry out, but, the **long term benefits** to the company will far out way the initial time taken and associated costs involved. As new information becomes available, this folder will be managed by and 'added to' by the Maintenance Manager with 'write access', all others will have 'read access' only.

- System overview.
- Control Philosophy - User Requirement Specification/ Functional Design Specification/ System Design Specification.
- Safety
- Piping & Instrumentation Diagrams.
- Process & Instrumentation Drawings.
- Process Flow Diagrams.
- Operation & Maintenance Manuals.

- 'As Built' Electrical Drawings.
- 'As Built' Mechanical drawings.
- Standard Operating Procedures/Work Instructions.
- Copies of Installation Qualifications/Operational Qualifications & Performance Qualifications.
- Routine/Preventative Maintenance Schedule
- Fault Finding
- Contact Information
- Spare Parts

Ensure a copy of the 'As built' set of electrical drawings resides in each machine's bespoke electrical panels across site at all times.

Once all the key engineering information is in place, the actual methods and systems of how to monitor and maintain all site equipment must be set up:

Work Instructions (W/I's):
Each piece of equipment will have O&M manuals supplied, which normally include installation, configuration, calibration, diagnostics, trouble shooting, service and maintenance instructions. Use these and the equipment manufacturers' expert advice to set up weekly/monthly/yearly work instructions to be 'carried out' and 'signed off' by onsite trained maintenance personnel for reference and service history purposes. The maintenance and servicing instructions make it possible for site personnel to implement preventative maintenance measures e.g. information is provided relating to periodical replacement of wearing parts in order to avoid damage to equipment.

N.B. The O&M manuals must be read before working with the equipment, for personal and system safety and for optimum equipment performance, make sure personnel thoroughly understand the contents before installing, using or maintaining the equipment.

Daily Logbooks:
Set up daily logbooks on key systems, again, using the equipment manufacturers' expertise to advise on what key operating parameters should be recorded and the acceptable tolerances the piece of equipment should be running between e.g. (7 Bar +/- 1 Bar).

Standard Operating Procedures (SOP's):
Engineering SOP's can range from 'Security and Access control' to 'Management of Site drawings'. SOP's may also involve the use of a 'specialist vendor/contractor' with the assistance of on-site trained personnel in carrying out the assigned tasks of the SOP itself e.g. Operability and maintainability of a Fire detection/sprinkler alarm system. It is vital the department manager ensures that training needs are identified and that training is performed before personnel carry out the task outlined within the scope of the SOP.

C. *Good Budget Strategy*

A good engineering budget is one of the most financially wise plans to have in organising and controlling financial resources as well as setting and realising goals. A budget allows the engineering department to figure out how much money it has to spend every month/year and where the department is spending it in a clear and documented fashion. As such, a budget is one of the most important steps to take toward maximising the power of the company's money.

Try to work from a 'zero based budget' platform. It is a technique of planning and decision-making which reverses the working process of traditional budgeting. In traditional incremental budgeting, departmental managers justify only increases over the previous year budget and what has been already spent is automatically sanctioned. No reference is made to the previous level of expenditure.

By contrast, in zero-based budgeting, every department function is reviewed comprehensively and all expenditures must be approved, rather than only increases. Zero-based budgeting requires the budget request be justified in complete detail by each division manager starting from the zero-base. The zero-base is indifferent to whether the total budget is increasing or decreasing.

Advantages of zero-based budgeting:

- Efficient allocation of resources, as it is based on needs and benefits.
- Drives managers to find cost effective ways to improve operations.
- Detects inflated budgets.
- Useful for service departments where the output is difficult to identify.
- Increases staff motivation by providing greater initiative and responsibility in decision-making.
- Increases communication and coordination within the organisation.
- Identifies and eliminates wasteful and obsolete operations.
- Identifies opportunities for outsourcing.
- Forces cost centre's to identify their mission and their relationship to overall goals.

D. Good Service Level Agreements (SLA's) with vendors and contractors.

A Service Level Agreement is a part of a service contract where the level of service is formally defined. It is a negotiated agreement between the company and the vendor/contractor and records a common understanding about services, priorities, responsibilities, guarantees, and warranties. Each area of service scope should have the "level of service" defined. The SLA may specify the levels of availability, serviceability, performance, operation or any other attributes of the service.

In advance of the commencement of any work on site, the proposed vendor/contractor must be firstly approved by site management; the following requirements must be fulfilled:

- Approval of all contractors/ service providers for use prior to coming on-site.
- Forwarding the Contract Company's Safety Statement & Method Statements/ Safety Data Sheets for the proposed work, if applicable to the site responsible person, who will obtain the necessary approvals.
- **Forwarding company details of the Contract Company's relevant insurances, tax certificate, Quality documents and employees' training records including Confined Space, M.E.W.P., Manual Handling, Forklift and all applicable training requirements.**
- Forwarding any other relevant Safety Training applicable to the work being proposed on site to the site responsible person.
- The Contract Company must also advise if any Subcontractors are to be used. If subcontractors are proposed, full details of insurance and environment, health and safety performance and competence must also be provided.

Only after the requested information is received and signed off by all the sites' responsible parties can the proposed Contractor/ Service Provider commence works on site. Contract Companies must only use persons whom they have assessed to be competent and suitably qualified to carry out the proposed work on site.

E. Set up a well stocked and efficient Engineering stores area

Companies are well aware of the cost benefits that can be attributed to a successful maintenance strategy. Part of that strategy will be to determine and assess the risks associated with the purchasing and storage of industrial spares and consumables based on their expected service life and their quality in performance versus their cost price. It is vital that personnel know and understand the products, services and materials that are suitable for purpose, work better, and last longer.

F. Set up a Maintenance Software System Package

A dedicated maintenance computer based software system or **Computerised Maintenance Management System** (CMMS) empowers the Engineering Department to manage its key functions and derive value for the business.
A maintenance software system allows the protection of a company's investment in assets and equipment. The system will help to manage maintenance, reduce downtime, control maintenance costs and minimise the investment in engineering spare parts. This type of system often becomes the showcase for maintenance departments during client and regulatory audits, it allows full visibility and traceability on all engineering and maintenance activities.

Key Benefits

- Manage both planned and unplanned maintenance work for both internal personnel and contractors.
- Improve the understanding of equipment behaviour, monitor recurring problems and costs.
- Gain a clear visibility on workload through effective resource planning.
- Analyse maintenance effectiveness through KPI Reporting.

- Prioritise maintenance activities through a simple graphic interface that highlights essential work.
- Create a knowledge base to identify reoccurring problems with equipment and to reduce time to repair.
- Reduce downtime through effective analysis of problems through a 'root cause analysis tool'.
- Manage a company's investment in engineering spares.
- Improve information quality with key information captured during fault resolution.
- Extend the life of assets through the effective management of preventative maintenance.
- Provide a unique historical record for all assets for both company and regulatory audits.

Key Features

- Full Asset Register storing key equipment information.
- Control Preventative Maintenance using both time and usage based intervals.
- Create planned and unplanned work and distribute work through a simple interface.
- Record completion of work along with time taken, downtime and defect causes.
- Create follow-up work orders to ensure that no activity is forgotten.
- Root Cause Analysis during Close Out.
- Fully Configurable System designed to meet a company's unique requirements.
- Downtime Analysis Reporting.
- Fault Analysis Reporting.
- Management Reporting Suite allowing interrogation of data.

Key Factors for Success

- A system owner responsible for the Maintenance System.
- Adoption of system by the Maintenance Department.
- Management buy-in.
- An Implementation approach.

- Spares Parts Management
- Purchasing.

G. Set up a Calibration Management Software System Package

This involves scheduling, documenting and controlling calibration activities:
Regular calibration and maintenance of field instrumentation is a must for many industries as they systematically prove the performance of their quality critical devices. The task of 'in situ' calibration itself is often challenging but add to this the need to schedule the activities, arrange with the production personnel to make plant available and then plan the resources and mobilise the technician... not to mention document the results is time consuming, costly and challenging.
Yet the key to ensuring the quality of the process whilst at the same time minimising costs for unnecessary maintenance are dependent upon the people who have to control this process.
Calibration Management Software is a high performance computer based software tool that meets all requirements regarding calibration management and implementation. Calibration Management Software allows a company to efficiently maintain and calibrate on-site instrumentation. It fulfils the high quality demanded of auditors and at the same time reduces complexity, time and costs associated with the management of the calibration activity.
Advantages:

- Controls the maintenance and calibration schedules, issuing work orders and reports as defined by maintenance personnel.
- Can attach important documents such as SOP's, Health and Safety sheets, loop diagrams to each tag.
- Incorporates audit trail and high security features.
- Up-gradable to full compliance of relevant regulations (Electronic Records and signature) with full validation services without loss of any historical data.

- Roaming laptop version is optionally available to allow technicians to work remotely on plant independent of the network.
- Produces management reports for cost analysis, historical data analysis, work due listings and overdue listings.
- Can print certificates related to discrete calibration or loop calibration.
- Is able to take into account unplanned requests for work.
- Full records sheet for each tag allowing all important data to be logged and recorded.
- Work areas can be arranged in plant locations and operating modules.
- E-mail control system which can be set to automatically alert pre-determined personnel of reports actions and deviations when action is necessary.
- Allows a company to control cost and review calibration schedules accordingly.

Lifecycle Management

Calibration Management Software will improve maintenance planning, improve the efficiency of a company's instrumentation department, reduce the cost of maintenance and at the same time satisfy the requirements of legislators and auditors.

H. Set up a site Energy Management Strategy.

Energy is a plant's biggest controllable operating cost. It is vital to keep that cost controlled in today's rising energy prices by optimising the efficiency of the plant and its associated machinery to obtain the maximum production output and to minimise downtime.

I. Emergency back up systems and contingency plans to ensure business continuity.

Preparation and time dedicated to reviewing and anticipating 'Worst case scenarios' by key knowledgeable users of all site equipment and activities cannot be underestimated. More importantly 'the layers of protection' needed to ensure the worst case scenarios never materialise, need to be put in place.

4
Safety

Safety first: All accidents, whether in industrial, farmland or domestic areas are preventable; most accidents are caused by failure to observe basic safety rules and precautions.

> *No job or deadline is so important that it may risk injury to people or damage to property or equipment due to work done in haste or improper installation.*

It is estimated in Europe, 15 - 20% of all workplace accidents are connected with maintenance and in a number of sectors over half of all accidents are maintenance related. 10 – 15% of fatal accidents at work can be attributed to maintenance operations. Therefore, it is vital that maintenance is carried out properly, taking into consideration workers' health and safety. In the construction industry worldwide, statistics show that a worker is killed on average every 10 minutes. That is 52,560 people per year.

During maintenance activities, direct contact between the worker and machine is inevitable; it is an activity where workers need to be in close contact with machinery and associated processes. Enforcement of safety and best engineering practices must be adhered to across a place of work at all times. This can be best achieved if there is a good workplace induction training program given to all personnel whether it is delivered 1:1 by a site EHS (Environmental, Health & Safety) contact, by PowerPoint presentation or video. This induction should include all site environmental, health and safety practices pertaining to the site's activities and will have at the end of the training session, a written assessment to prove competency.

These signed training documents will be reviewed and filed away by the site EHS contact, where they will be kept on record for future reference and also to flag when site induction training for personnel needs to reoccur (normally every 12 to 24 months).

> *It is vital a company is committed to carrying out all activities in a conscientious manner which protects the health and safety of its workers, the plant itself and the environment.*
>
> ***The first rule of safety is to work in a safe environment. 'Being safe' is in your hands.***

Toolbox talks provide a convenient and effective method of communicating and reinforcing the safety message throughout the workforce and when used properly, can significantly enhance the development of a safe working culture. The cost of implementing a regular toolbox talk system is minimal, 10-15 minutes a week. The benefits will include greater awareness, with the potential to reduce accident rates, and possibly even save a life.

Good communication is essential prior to carrying out any maintenance or installation work, the system owner i.e. area manager/production supervisor, must be informed in a timely manner by the person proposing to do the work of the impending task.

A '**Permit-to-Work**' form must be issued by the system owner prior to any non-routine work being carried out in their area. This may include a:

- Cold Work Permit
- Hot Work permit
- Line Break Permit
- Confined Space Entry Permit
- Excavation Permit

or any other work activity that the system owner may deem a Permit-to-Work necessary on safety grounds.

Permit-to-Work definition:
A Permit-to-Work system will identify hazards, specify control measures and describe work procedures. It is designed to plan and control all types of work which are potentially hazardous in nature and which could adversely affect the safety of personnel, the environment, or plant and equipment. It requires communication and coordination between site management, supervisory personnel, project/engineering personnel, operations personnel and contractors, to ensure that all work is completed in a safe and environmentally responsible manner.

The system owner may also need to be furnished with a well written, detailed '**Method Statement**' by the person proposing to do the installation of exactly what work is to be done and how it has been risk assessed. This method statement must leave no room for guesswork on the permit issuer's behalf.

Method Statement:
This formally describes the necessary 'Safe System of Work' prior to a task being undertaken.

Method Statement procedure must:

- Assess the hazards through risk assessment associated with the proposed system of work prior to that work commencing.
- Describe the controls necessary to eliminate identified hazards or to reduce the associated risk to an acceptable level.
- Describe the emergency measures required to mitigate the consequences of an associated emergency situation should it arise.
- Describe the training, roles, responsibilities and general welfare arrangements which must be in place prior to the work being undertaken.

When the method statement is completed and approved, it must be communicated to all stakeholders and personnel carrying out the task. Everyone involved must fully understand the proposed system of work, the associated hazards, the identified controls and the actions to be taken in the event of an emergency as documented on the method statement.

If a 'PTW' has been issued and work has commenced and an extra person needs to be added to complete the task and was not part of the original assigned team, their name must be added by the person who issued the permit. The permit issuer must make sure that the person is fully aware of all associated work and dangers pertaining to the ongoing task.

At all times one must be aware of the dangers around them. Cordon off work area with red and white barrier tape where maintenance/installation work needs to be done. Erect safety barriers, proper signage, i.e. maintenance work in progress signs and 'permit to work' forms visible for all to see, read and thus understand the nature of work being carried out.

Make sure that an appropriate **fire extinguisher** is selected, readily available and suitable for purpose. Be familiar with its operation, inspect the fire extinguisher and ensure to obey the recommendations on the instruction plate.

N.B. These procedures are put in place to protect other personnel in the immediate area and the person doing the work and never underestimate how essential this preparatory work is.

Always keep a work area clean and tidy; ensure items are not left thrown around to become tripping hazards. If oil, water etc. is spilled, ensure it is cleaned up immediately so it doesn't become a slip hazard. One is never too busy or has not the time, to keep a work area clean, tidy and hazard free. A person, who is in control of the task they are carrying out, always makes the time available for such activities.

At the end of each shift, ensure the work area is cleaned up and everything including tools, power tools etc. are restored to their proper designated place. A 'shadow board' for example, gives a clear indicator to personnel if a hand or power tool is missing. This makes it easy to know what goes where and have confidence that everything is where it should be.

The key point is that maintaining cleanliness should be part of the daily work routine, not an occasional activity initiated when things get too messy. Personnel must treat a work area, equipment, plant and material with respect. Keeping areas & equipment clean and tidy creates a better environment to work in; it also leaves a good impression on others, both inside and outside a facility.

N.B. All oils, sludge, slurries or drained off liquids are to be disposed of according to the pertinent environmental regulations.

Every person is ultimately responsible for their own safety. A person should take the extra 5 minutes before he/she carries out a task to satisfy themselves that they have all the properly maintained safety equipment as determined by the permit to work (PPE – Personal Protective Equipment) in place both on themselves (hard hat, safety glasses, hi-viz vest, safety boots, ear plugs, gloves, dust mask etc.) and in the immediate work area where the work is to be carried out.

Example of Template:

JOB TITLE	NAME OF EMPLOYEE	TRAINING REQUIREMENTS	PPE REQUIREMENTS
Supervisor	Joe Bloggs	A, B, C, F, H, I, J.	A, B, C, D, E, F, G
Fitter	John Smith	A, C, F, G, H, I.	A, B, C, D, E, F, G, J.
Fitter	Mike Mann	A, C, F, G, H, I.	A, B, C, D, E, F, G, J.
Fitter	Jim Carrig	A, C, F, G, H, I.	A, B, C, D, E, F, G, J.

TRAINING:
A=Safe Pass, **B**=First Aid, **C**=Manual Handling, **D**=MEWP, **E**=Confined Space, **F**=Abrasive Wheels, **G**=Working at Height, **H** = Site Induction others please specify, **I**= Chemical Handling, **J**= Rigger.

PPE:
A=Hard Hat, **B**=Safety Boots, **C**=Safety Glasses, **D**=High Visibility Vest, **E**=Ear Defenders, **F**=Gloves, **G**=Full Face Screen or safety goggles **H**=Chemical resistant gloves, **I**= Chemical Splash Suit, **J**= Full Face filtered mask for solvent vapour.

When any operation is likely to expose any employee on site to an average noise level of 80 dB(A) and above, an assessment shall be carried out by the EHS department and records maintained for company inspection. In such circumstances, the company must keep stocks of adequate ear defenders.

Lock Out/Tag Out System:
A company must ensure a **'Lock Out/Tag Out System'** is in place on site and **'LIVE BY IT'**. A person must never remove a lock that is not theirs from a piece of equipment, unless the other worker is completely unavailable and then, only remove the lock when the area supervisor and possibly the EHS officer, have verified that it is safe to do so and both have authorised the removal.

If a LO/TO system does not exist on site, instigate one. Insist that it is implemented. All affected employees must be trained in proper LO/TO procedures and be retrained regularly. It may someday save their life or the lives of their colleagues.

The "FATAL FIVE" Main Causes of LOCKOUT / TAGOUT Incidents:

(1) Failure to stop equipment

(2) Failure to disconnect from power source

(3) Failure to dissipate residual energy

(4) Accidental restarting of equipment

(5) Failure to clear work areas before restarting

LO/TO (Lock Out/Tag Out) Procedure Definition:
LO/TO is a safety procedure which is used in industry to ensure that equipment is properly shut off and not started up again prior to the completion of maintenance or servicing work. This procedure is put in place to outline practices, precautions, safety measures and hazards involved in the Isolation and Lock Out/Tag Out of any equipment or system that contains Hazardous Energy. The procedure of Lock Out/Tag Out is to be used to control Hazardous Energy. All isolations and Lock Out/Tag Outs of equipment and systems on-site shall be controlled with the Permit to Work System.

If a piece of equipment has been locked out/tagged out via an iso-lock. The LO/TO must never be removed unless all concerned parties remove their own personal locks and 'sign off' on the lock out/tag out. The permit issuer will then complete the final 'sign off' on the PTW once they have checked the equipment and are satisfied everything is in order both from a safety and operational perspective. The piece of equipment will then be returned to service.

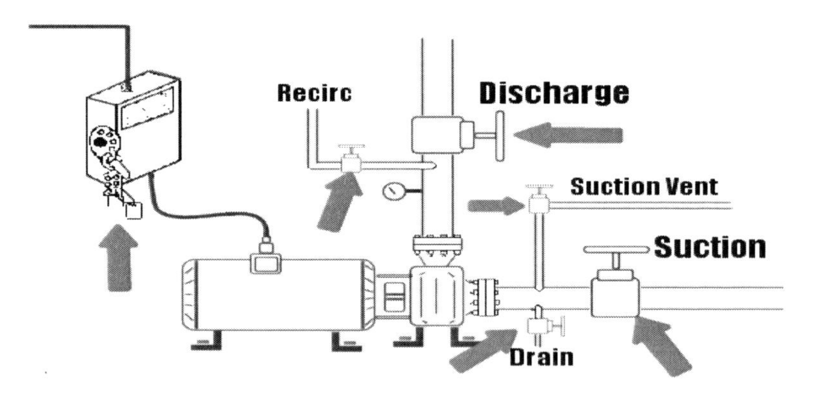

'Remember to lock and tag all hazardous energy sources.'

Communication is vital, especially during shift changes. A proper verbal and written handover should take place between outgoing and incoming shift personnel, pointing out any breakdowns, planned maintenance work done, any anomalies on equipment that may have taken place (no root cause found) during the previous shift and any work outstanding which is still 'locked out/tagged out' (i.e. the personal locks of the leaving shift must be replaced with the locks of the arriving shift). A written job card must be generated by the department affected. This job card/request may be written in paper hard copy or electronically in soft copy. The outgoing shift must make sure that the incoming shift understands the maintenance process and the associated hazards.

This job card should have all the vital information clearly written on it. It should include the name of the person who generated the job card, the status of work carried out and the name of the person who did the work. The date, time and hours spent doing the work must also be recorded.

When the work is successfully completed, all work carried out on the piece of equipment must be signed off (either hand written or electronically) by both the person who completed the work and handed back the piece of equipment 'fit for purpose' and the person who generated the job card, inspected the work carried out and accepted that the equipment is 'fit for purpose'. The completed job card must then be documented and available for future reference.

Having access to this documented information and being able to reference it easily in the event of an occurrence of a similar type of fault, can save hours of down time i.e. not having to go through what others have already experienced.

If a piece of equipment has been taken out of service and was **not** properly isolated mechanically or electrically (no 'lock out/ tag out' in place) or properly communicated through the proper channels, ask questions. Even if it means calling people who are off site, find out why it has been taken out of service without the correct procedure being adhered to.

In a suspicious situation, exhaust all avenues of communication/checks. Is it a shared utility? Has a person from another area or building turned the piece of equipment off without going through the proper procedures and channels? Try to find out why it was taken out of service in the first place and get agreement from peers/manager (if available) before deciding to put the piece of equipment 'back to service'.

The person, who did **not** isolate the equipment properly, must be part of the investigation and it should be explained to them that this type of work practice is unacceptable. They may need more training on such safety practices, pointing out also that their behaviour could endanger their work colleagues' lives and cause valuable downtime trying to establish why it was switched off in the first place.

N.B. When working on pressurised compressed air/water/steam systems, ensure that any pressure is isolated and safely vented to atmospheric pressure. Consider double isolation and the locking and labelling of closed valves. Do not assume the system has depressurised even when the pressure gauge reads zero.

Human error can prove very costly to both personnel and plant. 42% of **'all'** industrial accidents are the result of human error. Process and equipment failures account for the other 58%. Normally a critical chain of events occur that if overlooked may lead to disaster.
Sometimes personnel may have to make judgements and decisions based only on anecdotal information in a dangerous environment. They may have to interact with people when seeking more specific information. This leads to more critical time being lost, whilst at the same time trying to ensure proper procedures are being followed. A person's main focus must be to stop an

emerging abnormal situation from leading to a catastrophic accident.

If a piece of machinery is taken 'out of service' for a period of time and remains 'in situ', ensure it is still maintained even though it is not in use. Electrically and mechanically isolate, drain any liquids that have the potential to leak e.g. if a machine is left exposed to freezing air, water may freeze and split pipes.

If a piece of machinery is no longer in use and has been removed from site but all the various utility lines are still in place i.e. 'Live' electrical cables, compressed air, high pressure steam etc, remove them where possible back to source.

If this is not possible, clearly LO/TO all these utility lines and label each line. **Never underestimate the importance of these activities.** An improperly isolated utility line where electricity, compressed air, high pressure steam etc. may exist, can pose serious risk to both personnel and plant should it be 'turned on' or 'worked on' accidentally.

'NEVER TAKE A CHANCE'.

Never take any persons word that they have isolated a piece of equipment and it is safe to work on. **'Check it'.** Personnel must satisfy themselves, that everything is in order and that all proper safety and operational procedures have been adhered to.

Working at Heights:
If a person is working at heights, that person must always take extra precautions, wear a proper well fitted body safety harness and ensure they are clamped on/tied off to a solid strong permanent fixture that can take their weight in the event of a fall. Use MEWP's (Mobile elevated working platforms), properly erected 'tagged out' scaffolds by certified scaffolders etc. where possible.

When working at height:

1. *Always be aware of other personnel and their respective activities in the general work area. Have proper signage erected and cordon off the area that is being worked in with red and white barrier tape, ensure cordons do not hinder emergency egress. MEWP's are excellent vehicles to use when working at heights, remember also, they are big and powerful and can cause serious injury to personnel or damage to property if mishandled. Do not attempt to operate one unless certified to do so.*

2. *Be aware and ever vigilant of over head power lines when working with or moving MEWP's, scaffolds, ladders etc., striking a power line can cause serious injury and could be fatal.*

3. *Before deciding to ascend to the height that needs to be worked at, ensure that the ground the MEWP's, scaffolds, ladders are positioned on is solid, flat, clear of debris and can take the weight without fear of it collapsing and causing the MEWP, scaffold or ladder to topple over. Remember, even rolling over a stone on a roadway can be amplified many times when in an aerial platform at an elevated height and cause severe shaking. Whilst working within the basket of a MEWP, personnel must wear a full body safety harness with 'fall arrest retractable lanyard' securely attached to an anchor point within the basket.*

N.B.: When using an extension ladder, keep it the right distance from the wall, **use the 4-to-1 rule** i.e. for each 4 feet of distance between the ground and the upper point of contact (such as the wall or roof), move the base of the ladder out 1 foot. Place it on a level, firm surface and foot or tie at the bottom. Extend the ladder a minimum of 1.05m above the upper landing and tie off or anchor at the top, on the stiles not on the rungs.

N.B: A person is ultimately responsible for their own safety. ***One must never take a chance with their life.*** **All it takes is 10 seconds of thoughtlessness and lack of awareness e.g. trying to beat the traffic home, going up the outside of a scaffold instead of the safe designated way. Falling back onto solid ground, can leave a person paralysed for the rest of their life or possibly dead.**

Confined Space:

Confined spaces are significantly more hazardous than normal workplaces. **Extra precautions** must be taken when working in a confined space e.g. underground, vessel or tank etc. **Entry into confined spaces can be unforgiving as any errors can be fatal.** A seemingly insignificant error or oversight while working in a confined space can result in a tragic accident.

Safety must be based on:

1. A sound Management System.
2. Full understanding of all the hazards.
3. Thorough training.
4. Rigorous adherence to procedures.
5. Tested Emergency Procedures.

Is there another way of doing the task rather than the confined space option? The following factors must be considered and physically checked before vessel entry is to be carried out:

- Previous contents
- Residues
- Contamination
- Oxygen deficiency and oxygen enrichment
- Machinery
- Physical dimensions
- Work to be carried out:
  cleaning chemicals
  sources of ignition
- Dangers from outside

Could hazardous vapours or gases be formed by the work being done?

When working in confined spaces the following points must be implemented/addressed:

- Issue of a permit to work
- Work scope and method
- Nominated Supervisor
- Tally man / Attendant
- Rescue procedures and equipment
- Training (mandatory)
- Tools and equipment to be used, including low voltage or pneumatic
- Lighting requirements, including standby/emergency
- Explosion proof fittings
- Ventilation
- Access
- Bonding to prevent both electrical shock and static discharge
- Work cycles, to reduce risk of heat exhaustion
- Fire safety and extinguisher requirements.

Personnel shall not enter or commence work in any excavation, tank, vessel, pipe or chamber or other enclosed space, until a valid permit to work (confined space permit) has been issued by the Issuing Authority. Where operations result in a dangerous atmosphere arising during the monitoring of the work activity, the PTW issuing authority must be informed and all personnel removed from the area.

No new activity shall be introduced into a confined space without the permission and signed approval of the PTW issuing authority. Whilst work is ongoing within a confined space, the company will be required to provide a properly trained standby/attendant person. All personnel who have to enter confined spaces must have undertaken the training appropriate to this task.

No site personnel can enter a confined space to carry out a work activity in that confined space unless there is, in respect of that confined space, suitable and sufficient arrangements for the rescue of persons in the event of an emergency. A confined space register which includes an assessment of known and potential hazards for all 'Confined Spaces' on site must be provided. Detailed risk assessments, which define the manner in which the risks identified for the confined space and the work to be carried out within the confined space are to be controlled must also be provided. A standard test for vessel entry would be, e.g.:

Atmospheric Test (pre-entry):
A test performed by a competent person to determine the suitability of the atmosphere for breathing by persons and a safe environment with respect to the flammable, combustible and explosive potential prior to entry into the confined space. Results of this test will be recorded on the Confined Space Entry Permit / Work Permit prior to the work being carried out and further tests checked and written recordings made whilst the work is ongoing.

Before entering any vessel, tank, silo etc., a **'Confined Space entry permit'** must be completed and fully signed off by the system owner and all personnel involved in the work. All involved must make sure any associated gaseous/electrical/mechanical apparatus attached to the vessel is totally identified, isolated, removed/spaded/blanked off, locked off with each individual having their own respective key for any 'isolock' installed.

Both the person entering the confined space and the second person acting as a 'lifeline' must ensure there is proper access/egress maintained constantly. The person entering the confined space must at least be watched at all times from outside and all appropriate precautions must be taken to ensure that they can be assisted effectively and immediately. Ensure also, a constant means of communication exists between the two parties while the confined space entry permit is active.

When working alone in remote areas, personnel must **always** inform a **site designate** of their location and the anticipated duration of their task before commencing work. If a person does not and something unforeseen happens to them, they may not be missed for some time.

Tag System

The air a person breathes in the Earth's atmosphere is comprised of approximately 78% Nitrogen, 20.9 % Oxygen and the other 1% or so made up of special gases. If the oxygen % level drops, it can lead to a person almost instantly collapsing or be fatal.

N.B. Never make even a partial entry into a confined space unless it has been tested and proven safe. Never even consider the possibility of holding one's breath and entering the confined space.

If a person finds themselves in the unfortunate circumstances where a work mate has collapsed in a silo or tank and they are the standby person. **They must never go in after them,** as they you may suffer the same fate, instead call for emergency assistance and try to winch them out safely.

***Remember,* by just putting your head into an open pipe or tank, you can be overcome and collapse in seconds if the oxygen content is in any way diminished.**

Think and be aware, before entering into a confined space. Personnel must not take unnecessary risks and put their life or the life of their colleagues in danger just because they want to finish the job quickly and get home to see the game on TV.

The person entering the confined space must wear a full body safety harness with 'fall arrest retractable lanyard' securely attached to an anchor point e.g. Davit arm, mobile tripod with proper certified winching gear attached to the fixture so a person can be removed from the confined space easily in the event of them being injured or collapsing. They must also wear all other appropriate PPE as dictated by the PTW.

Excavation and Openings Works:

- Prior to the start of any excavation on site, engineering must be consulted and the presence of overhead and buried services shall be checked. Where "live" services are present, hand excavation must be carried out until the location of the service has been identified, recorded and made safe.

- Personnel must erect suitable solid edge protection (i.e. double handrails) around excavations or openings. During the hours of darkness, any excavations, openings or obstructions near or on roadways and walkways must be indicated by a sufficient number of warning lamps.

- The sides of all excavations should be properly shored, battered or stepped to prevent collapse. No excavation work shall commence unless there are adequate resources present to ensure the stability of the excavation. Excavations shall be inspected prior to commencement, or re-commencement of the work to ensure the excavation is still in a safe condition.

Keep safety equipment certified and in good condition at all times. If damaged, worn out or has passed the manufacturers date of safety guarantee (e.g. check the safety guarantee date stamped on a safety helmet) it must be replaced, not because it has to be done and it is site regulations, but, because it is good safety practice and it may someday save a person from severe injury or death.

N.B. Personnel must observe all valid health & safety provisions and regulations pertaining to their country.

Fundamental criteria that must be addressed and implemented when designing and working with machinery to ensure safe, environmentally friendly, reliable and cost efficient operation should be:

1. To provide properly designed, fully documented safe automation technology.
2. To ensure 'fit for purpose' reliable electrical, mechanical, control and instrumentation equipment is installed and properly maintained.
3. To ensure relevant specific real time information with proper context is being generated and is user friendly.
4. To ensure specific knowledge, information and skills are communicated clearly and associated abilities attained in a way that personnel can understand and apply to their specific work.
5. To ensure personnel have the ability to disseminate and understand real time information easily and act accordingly and also have the ability to make multiple decisions at any point in time.

When considering safety in relation to a piece of equipment:

- Ensure that proper emergency safety equipment is installed to mitigate the consequences of an emergency event.

- The piece of equipment must be capable of operating to its designed specifications and conditions with an appropriate safety factor.

- The piece of equipment must be protected from conditions that might occur outside of its operating conditions or capabilities and within safe limits of its operation e.g. an air compressor has to supply compressed air to a production process. The compressor itself has been **incorrectly** installed in an area where sub zero temperature conditions may exist from time to time. The compressor's minimum protection start up temperature of > 2°C zero was not taken into consideration. Compressor will not start below 2°C, thus not allowing production processes to start up.

- The piece of equipment must have appropriate safety features that protect personnel while installing or maintaining it. Ensure hands and clothing are protected. Do not wear loose clothing (ties, scarves etc.) or jewellery and tie back long hair. **All of these could become entangled in a pulley or be dragged into a rotating mechanism.** The equipment must be installed in a manner that allows for these safety requirements. Stay clear of all rotating parts and of all moving parts, ensure safety guards are in place at all times.

- Chips or debris may fly off objects when they are being struck (safety glasses must be worn). Before objects are struck, try to ensure no other personnel will be struck and possibly injured by flying debris.

- **Ensure to support any piece of equipment properly when working on or beneath it** i.e. any type of machinery, car, truck, tractor, trailer etc. A person **must not under any circumstances**, depend solely on a car/truck jack to support the vehicles weight if they have to physically go underneath it to investigate or fix something. It is vital to use a secondary 'fail safe' support e.g. stacked concrete blocks or an RSJ properly positioned under a 'weight bearing' part of its frame which can take the vehicles weight if the car/truck jack fails or falls to one side while they're still underneath it. Ensure also to brace the wheels of the vehicle with a block of wood/concrete to prevent it from rolling while working on it.

- Personnel **must not** mount or dismount a piece of equipment other than at designated locations that have handholds or steps. They **must not** stand on components mounted on the equipment e.g. solenoid valves, probes, filters etc. to elevate themselves to a higher position. Use an adequate ladder or a work platform. Secure the climbing equipment so that it does not move.

- Do not carry tools or supplies when mounting/dismounting a piece of equipment. Use a hand line to raise and lower hand tools or supplies.

- Death may result if safety mechanisms are removed or 'nonfunctional'. Removed safety mechanisms (e.g. for cleaning, maintenance and repair work) may pose a health hazard or cause a fatal accident. Replace the removed safety mechanisms immediately on completion of the assigned work and test them if necessary. Check the safety mechanisms on a regular basis.

- Equipment that has been in service may contain trapped pressure and/or residual media even after washing. Opening and/or disassembling of machine components should **only be carried out by trained personnel** in strict observance of the manufacturers' instructions. The same applies to the safety precautions which need to be taken when handling the residual media itself.

N.B. Do not use emergency fire alarm 'break glass', 'emergency stop' buttons, electrical isolating switches etc. as coat hooks for a jacket, sweater or any other piece of clothing. Not alone is it obstructing a piece of emergency equipment but it is also blinding other people to its actual location in the event of an actual emergency. A person could also inadvertently activate the device by simply leaning against the piece of clothing. If an emergency stop is activated or any other piece of emergency equipment in 'error', ensure it is reported to the area manager.

Set up an Environmental, Health & Safety (EHS) Systems Software Package

A dedicated systems software package is a computer based tool which helps a company to easily manage and track their key functions on Environmental, Health & Safety issues i.e.:

- Accidents
- Incidents
- Near misses
- Environmental Impacts
- Non Conformance
- Safety risk assessments
- CAPA's
- Monitoring
- Objectives, targets and plans

The system allows full visibility and traceability on all company EHS activities.

Key Benefits

- Analysis of system effectiveness through KPI Reporting.
- Creates a knowledge base to identify reoccurring problems.
- Reduce EHS issues through effective analysis.
- Improve information quality with key information captured during issue resolution.
- Provides a unique historical record of all EHS issues for both company and regulatory audits.

Key Features:

- Record of all Incidents/Accidents.
- Root Cause Analysis reporting along with the time taken to satisfactorily complete investigation and close out.
- Creates CAPA's.
- Fully Configurable System designed to meet a company's unique requirements.
- Management Reporting Suite allowing interrogation of data.

Emergency Response Preparation:

Never become complacent about the hazards in a workplace.

Emergency Phone Numbers:

Ensure the emergency medical numbers, especially for on site **'First Aid'** personnel, are available and easily accessible, as well as local hospitals and doctors.

Role play and preparatory training on site for accident and incident scenarios should be mandatory. Knowing who to call (e.g. security, shift manager or main site utility providers i.e. electricity, gas, water, emergency 'on call' phone numbers etc.), what to do (e.g. use a fire extinguisher) and how to escalate an emergency situation (e.g. activate site fire alarm) must be known by all personnel in the workplace. These escalation procedures should also be part and parcel of any proposed works' method statement.

It is vital personnel implement all provisions valid for their country and pertaining to the opening and repairing of electrical/mechanical devices. All instructions regarding the repairing of electrical and mechanical devices must be adhered to without fail.

Personnel must absolutely and without fail, read and understand the relevant piece of equipments' O&M manuals before carrying out its instructions. Only trained qualified personnel can carry out work on a piece of equipment. The detailed maintenance instructions and safety precautions, as per the Technical Operation and Maintenance Manual of any piece of equipment, must be followed at all times.

Never be tempted into cutting corners on safety. Ensure proper maintenance intervals are adhered to as recommended by the manufacturers of the equipment supplied to the company. **Do not take shortcuts**; this may lead to failures later on. It can also leave the equipment in an unsafe condition, putting other people's lives at risk.

N.B. When working on hazardous liquids or gases in a pipeline, consider what is in the pipeline or what may have been in the pipeline at some previous time.

Consider:

1. Flammable materials.
2. Substances hazardous to health.
3. Extremes of temperature.

Incident/Accident:

If an incident/accident has taken place, the investigation must begin immediately. Capturing as much information as possible in the first 30 minutes is vital.

1. Determine if the incident/accident is under-control.
2. Determine if the incident/accident has had or may have any potential knock on impacts.
3. **Cordon off the area.**
4. Determine the resources that are required to deal with any or all the issues.
5. If the incident/accident is of such a nature that it has, or may have, off site implications, it may be necessary to gather an incident management team on site.
6. **Ensure only 'authorised personnel' enter the 'cordoned off' area to gather the evidence/facts.**
7. **If a number of people witnessed it, ask everyone involved to immediately write down and document exactly what they remember happened and what they were doing at the time of the incident/accident e.g. operating an agitator via a PC screen or manually opening a hand valve allowing liquid into a tank.**
8. **Take photographs and measurements. Record anything out of the ordinary as soon as possible. This type of recorded information can prove vital during the subsequent investigation and may actually help find the 'root cause' of what caused the incident/accident in the first place.**
9. **Treat a person's comments with courtesy and respect at all times.**

Whatever work is carried out, **'Do it right first time'**, regardless of the time pressures. Do not put a piece of equipment 'back to service' unless 100% satisfied that it is safe to use, fit for purpose and fully operational.

N.B. To be a focal point of an accident/incident investigation because of something you did can be a very lonely place, especially if people are injured or dead, environmental damage has occurred and serious monetary loss has been incurred by the company. Making excuses like 'I didn't think', 'I was watching the clock', 'I was under pressure to get the job done' while someone lies dead in a morgue because of your actions will be 'cold comfort'.

Safety Overview

- **Safety first:** All accidents are preventable.

- The first rule of safety is to work in a safe environment.

- No job or deadline is so important that it may risk injury to people or damage to property or equipment due to work done in haste and equipment not properly installed.

- Before starting any work, you must satisfy yourself, that everything is in order and that all proper safety and operational procedures have been adhered to.

- Enforcement of safety and best engineering practices must be adhered to across the site at all times.

- Every person is ultimately responsible for their own safety.

- Human error can prove very costly to both personnel and plant; 42% of all industrial accidents are the result of human error; process and equipment failures account for the other 58%.

- Communication is vital; a proper verbal and written handover of all relevant site activities should take place between outgoing and incoming shift personnel.

- A company must ensure a **'Lock Out/Tag Out System'** is in place on site and **'LIVE BY IT'**. A person must never remove a lock that is not theirs from a piece of equipment, unless the other worker is completely unavailable and then, only remove the lock when the area supervisor and possibly the EHS officer, have verified that it is safe to do so and both have authorised the removal.

5

Good Alarm Management

Most control systems that monitor, control and operate equipment are equipped with a series of monitoring devices to ensure reliable operation. If any irregularities occur, an alarm is triggered and the equipment carries out the necessary switching steps automatically in order to avoid hazardous situations. The alarms associated with these monitoring devices usually range from having no direct influence on the system to its complete shutdown. These alarm set points are normally sub grouped in order of there criticality and indicated in the alarm text with colour coding on a PC/HMI screen e.g.

Alarm Group 1
– Pre alarm without direct influence on the system.

Alarm Group 2
– Partial shutdown of the system. (Hold mode)

Alarm Group 3
- Complete shutdown of the system.

Alarm grouping and filtering facilities must be provided to allow for alarm segregation between plant process trains and equipment.

The ultimate objective is to prevent, or at least minimise, physical and economic loss through automatic or manual intervention in response to the condition that was alarmed.

N.B. A company must revise its procedures for periodic testing and verification of its 'back-up control systems' to assure they will function properly at all times and alarm accordingly.

The fundamental purpose of alarm annunciation is to alert personnel to abnormal operating conditions.

Good Alarm Management principles are vital in every industry and must be properly implemented; procedures put in place and adhered to, practiced and properly escalated according to the criticality of the activated alarm. It is vital that site management adopt these principles. There are sometimes too many alarms annunciated in a plant because of poorly designed alarm strategies (possibly back at the URS design stage) improperly set alarm points, ineffective annunciation, unclear alarm messages, etc. Design teams must differentiate between alarm management and alarm flooding and **try to avoid alarm saturation i.e.**

1. Massive quantities of data arriving on the PC's alarm banner that are meaningless and without proper context or explanation.
2. Alarm saturation makes it difficult to analyse which alarms need immediate attention.
3. Alarm saturation makes it difficult to make important informed decisions and take swift action.

It is vital to reduce **'alarm fatigue'** so personnel know that any alarms generated by the system are serious enough to demand their intervention. If the resultant flood of alarms becomes too much to comprehend, then the basic alarm management has failed as a system.

The speed and accuracy with which critical alarms can be identified and require immediate attention will determine how effective an alarm management system is and how quickly it can be responded to.

It must be decided by operations/engineering (mainly operations) which alarms need more attention than others. As well as the PC screen icon flashing, an audible siren may need to be installed to alert personnel of the severity of such a critical problem. It is only by eliminating extraneous alarms that only need to be recorded and not have an audible alarm attached to them and providing better recognition of critical problems that a faster, more accurate response will occur. Alarm suppression facilities should also be provided for and prioritised accordingly.

Insist during the actual installation of equipment that as much information regarding each system can be put up on the PC screen to assist personnel in diagnosing a fault. Install 'shortcut icons' in a corner of the PC screen which will take personnel directly to view the piece of faulty equipment's documents, e.g.

1. SDS (Software Design Specification).
2. Machine's P&ID.
3. HVAC P&ID.
4. Parameters e.g. set points, volumes, fan speed set point etc.

These helpful 'shortcut icons' and the stored relevant information attached to them, will save a lot of plant downtime as the information required is easily accessible.

Proper alarm management and annunciation is vital in alerting all concerned to abnormal operating situations in areas such as:

1. Personnel.
2. Properly escalating a fault/issue and the response to same.
3. Environmental safety.
4. Equipment integrity.
5. Product quality control.

6

Training

Personnel training remains one of the most cost effective ways to cut costs and improve the efficiency of a company's operations. Investing in training to stay abreast of industry's latest techniques and requirements will play a key role in the success of any company.

From the most basic to the most advanced, anything a person does is an improvable skill. Improvement comes from learning. The quality of how a person approaches their work, the quality of how they actually do it and the continual improvement and innovation of both.

Improvement issues and the changes required need to be identified. We are taught new information everyday in classrooms, the workplace or at home, **but we are not always taught how to 'think'**.

Staying focussed, selecting the correct information that needs to be learned and communicated to personnel as to its relevance, will be key. If all work activities throughout the company are approached with the same mindset and **attitude**, information will be processed more deeply and be retained longer. Constantly strive to do things better and take responsibility for any actions undertaken.

3 Key themes for the information age of the 21st century are:

1. Knowledge.
2. Lifelong learning.
3. Education.

The most valuable asset any company has is its people. They are the driving force behind its success and the key to its future, but, if these people lack essential knowledge and competencies, their value and the company strength will be seriously diminished.

For a company to remain competitive, it will need to keep abreast of developments in technology in order to exploit the benefits offered by new technology. Investment in new technology on the factory floor will be seriously undermined without the corresponding investment in the technical awareness of personnel. **Such awareness comes from training.**

Investing in the correct training now will give a business the leading edge it needs to compete in today's market place. By raising the level of technical understanding, people will be empowered to play a more important and active role. A company's success can be assured through increased efficiency and reduced downtime.

Technical training should encompass theoretical knowledge, system design, system building, evaluation of system performance, fault finding, maintenance, documentation and safety of as many engineering disciplines as possible e.g. electronic, electrical, mechanical, hydraulic.

Employers are looking for versatility in employees who are multidisciplined and will cover all aspects of machine maintenance, remember **'Job security depends on flexibility'**. Just because personnel haven't been taught how to work with for e.g. 'Pneumatic systems' in their previous training doesn't mean they can't learn it and become very proficient in the discipline.
'Attitude' is the key word, a person must not wait around for someone else to progress their career, they must make it happen themselves. Use the internet, there are excellent 'online' free training courses being provided by the World market leaders websites in electrical control, hydraulic systems, steam systems

etc. Personnel must stay 'up to date' and ensure any relevant training received pertinent to their job doesn't lapse. For example, reviewing old notes and training modules to refresh one's mind on a subject.

Personnel training should be based on:

- **Skills** – Have the proper skills been attained required to carry out proper maintenance on specific equipment?

- **Competency** – Have they the abilities to satisfactorily perform the required functions of their job?

- **Knowledge** – Have they a clear understanding of the systems operations and capabilities i.e. all the knowledge needed to run the plant safely, efficiently and confidently.

The term **training** refers to the acquisition of knowledge, skills, and competencies as a result of the teaching of practical skills and knowledge that relate to specific useful competencies. A person must recognise the need to continue training beyond initial qualifications and to continue to maintain, upgrade and update skills throughout their working life. Remember, personal development is about enhancing skills and knowledge.

Training should ensure that personnel have good knowledge of the equipment documentation provided by the vendor and are capable of operating and maintaining the equipment without danger to them or the equipment without the vendor's supervision. This training normally takes place before and during the installation and commissioning phases. Ensure the vendor provides a good training plan to encompass all relevant personnel that actually need the training. The more onsite people who are trained, the better chance of resolving a major fault if one develops through the accumulated knowledge of the collective.

Training Environment

Ensure the training equipment is as similar as possible to the 'real life'equipment. Providing training on 'simulation systems and models' as well as procedures facilitates greater personnel performance and retention.

Frequency of Training

Cognitive tasks normally decay faster than physical tasks, to prevent this happening:

- More frequent refresher training is needed for those types of tasks to avoid loss of knowledge.
- Use bursts of 'knowledge bites' training to maintain that knowledge.

It is vital to the success of the company that all knowledge does not reside with one person. This type of management is narrow minded and will leave the company badly exposed if something happens to this person. A company's training vision must include as many people as possible to never allow such an issue to arise. Every eventuality cannot be catered for, this fact must be recognised.

Written and verbal assessments are vital to prove competencies. These assessments must be given and corrected only by the experienced systems experts. Whether it is given by the vendor of the equipment or an onsite specialist, the trainee can only be passed competent by the system experts. All this training must be documented, auditable and flagged when refresher training is required. Personnel need to be 'up skilled', if the piece of equipment has been upgraded. A company must keep up to date with the latest upgraded training versions of equipment supplied by vendors to the plant.

If personnel feel the training they received on a piece of equipment is not adequate, **'They must say so'**. The onus is both on the company and the person being trained to ensure proper training is given and received. If the system vendors' training is not good enough, the vendor must be written to or told verbally. The depth of knowledge of experienced trainers is key. Trainers must be in touch with 'day to day' issues and their solutions and also the practical aspects of the person being trained (e.g. roles and responsibilities). Any course given should be updated in-line with the latest regulations, technical developments and customer feedback.

Training starts with the individual and their personal development.

- What do I need to improve?
- How do I go about improving my skills and abilities?
- Do I have a plan and deadlines in place in trying to achieve the goals I set for myself?

These fundamental questions must be asked and answered honestly. It has to start with the person themselves and then ask others for their professional advice and opinions to help them gauge how other people perceive them, regarding how they go about doing their 'day to day' job.

A person must **'Train by doing'** and don't allow their education to lapse, keep up to date with the latest skills and knowledge required for their current role. What a person knows is crucial, it shows to their current employer that they are a life long learner by gaining advanced skills and expanding their general business knowledge. Potential future employers will also appreciate such initiatives.

If working with a company who sets aside a good training budget for its employees, ask them to pay for relevant courses. Point out the reasons for such courses and the benefits both financially and operationally to the plant as a whole. It is vital a company maximises the benefits of training. It is imperative to determine not only those technologies in which staff have a knowledge deficit, but it must also identify what level of training is required for each person to successfully and efficiently perform her or his role within the plant. Efficient learning is about being able to put newly acquired knowledge and skills into practice, ensuring that skills acquired in the class room transfer readily into the work place.

Recognise strengths and work on weaknesses. Never be afraid of asking questions or taking additional training.

Don't vegetate. Go on as many engineering courses, seminars etc. as possible. Meet other people and find out what their companies are doing. It is also a very good way of networking, setting up contacts and being able to ask other's for their expertise, experience and advice on universal issues e.g. vendor selection, equipment selection, value for money etc.

Each employee in the company should have a designated folder assigned to them with all their relevant personal details, photo etc. along with their original copies of their initial qualifications. All subsequent 'in house' or 'off site' training with all certificates of courses attended should be tracked via a 'Training Matrix' i.e. SOP's, Training Methods, Internal training and new skills acquired. This folder must be held by the HR department for future reference and updated as an employee's future personal circumstances change and new qualifications are attained.

Improvements from learning mean doing things better, faster and cheaper.

Setting goals and having the vision, confidence, persistence and perseverance to pursue and achieve them will be part and parcel of any successful career.

7

Competency

Formal measures of competence are increasingly being demanded. Competence is the key to ensuring that both the company and staff achieve the highest possible level of productivity, it is a standardised requirement for personnel to properly perform a specific job. It encompasses a combination of knowledge, skills and behaviour utilised to improve performance. More generally, competence is the state or quality of being adequately qualified or having the ability to perform a specific role.

Competencies depend on the skills and knowledge of individuals. Ensure with any task a person carries out, that they are always within their ***'envelope of competence'***. Staff competencies must be addressed with respect to the competencies required for a company to optimise performance and help meet its perceived challenges.

A person is deemed competent if the skills, abilities, and knowledge they possess enable them to perform effective action within a certain workplace environment.

Goal: ***To increase the level of engineering competency, to measure and improve the knowledge and skills of all engineering practitioners.***

Definitions:

- Competency is the ability to satisfactorily perform the required functions of a job, usually defined in terms of tasks or levels of skill for a specific job.
- Certified competence is a credential or certification that an individual has the basic skills, knowledge and experience to carry out tasks on a piece of equipment.

The progress of technology remains an ongoing challenge as any sizeable piece of equipment is now made up of electrical, electronic, mechanical, hydraulic and pneumatic modules operated via a PLC with a HMI attached. There is a constant flow of new technologies in the engineering field requiring new knowledge and skills. New technologies arrive, requiring competency in an ever increasing range of technologies. An estimate is that it takes **five to seven years** of 'on the job' experience to develop a contributing engineering practitioner.

What must be identified are the **competency gaps** i.e. skill or knowledge areas in which the ability of the individual or organisation might not be sufficient to meet the business needs. There are very reputable companies who will work with a business as a training partner. They will review the current competency levels of staff against those required. They will then help identify deficits in the skills and competencies required of personnel and can develop customised training solutions to bridge this shortfall ensuring a company achieves its strategic training targets.

Competent employees will be the key to any successful company both now and in the future, in other words '**Doing things smarter and better**'.

Always try to be
a 'Solution Provider
not an Obstacle Generator',

Senior management like problem solvers, not problem seekers.

8

Instrument Calibration and Validation

Calibration is a process of comparing the accuracy of an 'instrument reading' to known standards. It is a demonstration that a particular instrument or device produces results within specified limits by comparison with those produced by a reference or traceable standard over an appropriate range of measurement.

Fully documented, properly indexed, stored calibration reports and calibration certificates with individual serial numbers for all relevant instruments, are essential in ensuring the smooth running of a plant in a safe and cost effective manner. All instruments shall be factory calibrated and certificates of calibration shall be provided with each device. This usually involves the instrument vendor commissioning the installed instrument and issuing a field certificate prior to system handover e.g. a radar liquid level detector being installed in an agitated tank.

All calibration certificates must be traceable to the National Certifying body. For proper operation and optimal performance in a process, each element of a measurement system must be carefully chosen. It is vital to demonstrate that each component of the measuring system is functioning to specification and is fully auditable. If any component of the measurement system does not meet this requirement, out of specification product may be produced leading to costly rework or worse, may lead to an accident.

Use of certified, traceable electrical simulators and standards to verify an instrument is processing input signals correctly is mandatory.

E.g. a temperature sensor's ability to generate the correct output signal so that an Electronic head transmitter/PLC/SCADA system can interpret this signal and ensure that it corresponds accurately to a temperature reading.

By using a calibrated resistance decade box and injecting a known resistance into the instrument, this pre-set resistance should then correspond precisely to a correct output signal.

The temperature sensor's transmitter probe is programmed and set up to read as follows:

Temperature range:
(0 – 100°C)

Standard corresponding resistance input:
(100 ohms – 138.51 ohms)

Standard corresponding current output:
(4 mA – 20 mA)

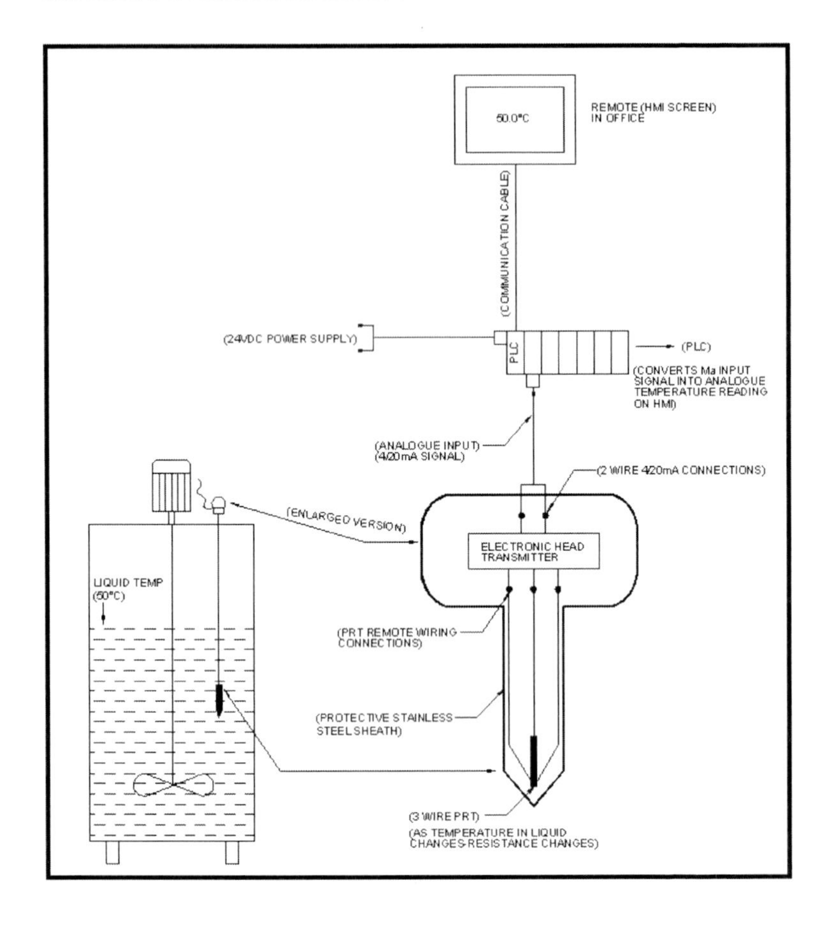

The **PRT (Platinum Resistance Thermometer)** itself and its associated wiring are normally encapsulated and protected by a stainless steel covering. The PRT is positioned at the tip of the measuring probe, is about 25 mm in length and is connected to the head of the probe with either 2 or 3 wires (lengths of the probes can vary from 10 cm to > 3 metres).This type of temperature sensor exploits the predictable change in electrical resistance in the platinum with changing temperature. The PRT provides a known resistance as the temperature fluctuates. The electronic transmitter within the head of the temperature instrument itself is programmed to understand this fluctuating resistance and subsequently converts to a milliamp signal.

To test functionality:

1. Disconnect the wiring from the 'PRT input connections' of the temperature transmitter electronic head, push them to one side.

2. Proceed to connect the wires from the resistance decade box directly into the input connections of the temperature transmitter electronic head.

3. Inject a resistance of 119.40 Ohms. A corresponding reading of 12 mA should now be measured on the output connections of the instrument using an ammeter connected in series with the output wiring going to a PLC. This 12 mA signal is then converted by the PLC to read 50°C on the HMI screen via a communication cable.

4. 5 known resistances of between 100 Ohms – 138.51 Ohms are inputted into the instrument which gives an equivalent electrical output signal of 4 - 20 mA which should then show a temperature range of 0 – 100°C, where:

Input: Resistance 100.00 Ohms = Output of 4 mA = 0°C
Input: Resistance 109.73 Ohms = Output of 8 mA = 25°C
Input: Resistance 119.40 Ohms = Output of 12 mA = 50°C
Input: Resistance 128.99 Ohms = Output of 16 mA = 75°C
Input: Resistance 138.51 Ohms = Output of 20 mA = 100°C

N.B. Once the test is complete, reconnect all wiring exactly the way it was disconnected, observing colour coding and polarity at all times.

A **thermocouple** may also be used instead of a PRT. A thermocouple is a junction between two dissimilar metals that produces a minute voltage in proportion to its temperature. Thermocouples for practical measurement of temperature are junctions of specific alloys which have a predictable and repeatable relationship between temperature and voltage. They are a widely used type of temperature sensor for measurement and control.

A comparison of a temperature instruments performance can also be made with a certified reference thermometer or a temperature controlled heating/cooling bath. This actually proves the instrument itself and all its associated cabling without physically disconnecting any wiring i.e. the certified reference thermometer is put into the tank of liquid and the reading is then cross referenced with the digital temperature readout on the HMI screen. The temperature measuring probe can also be removed from the process and put directly into a calibrated temperature controlled heating/cooling bath where known temperatures are set. Checks are then made to see are the readings set on the calibrated bath (e.g. 50°C) corresponding directly to what the digital temperature readout is reading on the HMI screen itself.

Calibrations intervals should be performed at least once a year or as specified by vendor recommendation, but for **critical applications** a greater calibration frequency may be required to provide the required confidence in the process instrument.

Instrument Validation means establishing documented evidence which provides a high degree of assurance that a specific instrument will consistently produce accurate readings, meeting its predetermined specifications and ensures that the specific instrument is working correctly and fit for purpose.

Criticality Assessment Team Assessments should also be carried out on instrumentation and equipment that monitors and controls equipment on plant. A critical assessment of all equipment should be made, based on the criticality of the piece of equipment. Regular maintenance and calibration intervals must be maintained and logged to ensure equipment remains safe, compliant and fully operational at all times.

(CAT) Process

This is conducted, system by system for each plant or area. A complete instrument list is generated by reviewing approved P&ID's and/ or 'walking down' each instrument in the field. A cross functional team comprising of the System Owners, Engineering, Quality and EHS categorise each instrument as being one of the following three categories.

- GMP Critical
- EHS Critical
- Non-Critical

N.B. Failure of a critical instrument or late execution of a critical calibration may require a deviation or safety report to be raised and an investigation to be carried out as per a company's SOP system.

GMP (Good Manufacturing Practice) Assessment

GMP critical instruments are assessed as per the criteria defined in an SOP e.g. Quality Impact Assessment Procedure.

Instruments are defined as being "GMP Critical" if they were judged to be "Operationally Critical". To be operationally critical, each instrument is critiqued against the following questions:

1. Is the instrument used to demonstrate compliance with a registered process?
2. Has the normal operation or control of the instrument a direct effect on product quality?
3. Does failure or alarm of the component have a direct effect on product quality or efficacy?
4. Is information from the instrument recorded as part of the batch record, lot release date or other GMP related documentation?

5. Does the instrument control a critical process element that may affect product quality without independent verification of the control system performance?
6. Is the instrument used to create or preserve a critical status of a system?

If the answer to any of the above questions is "Yes" and there are no other parallel or downstream components that would detect failure of the instrument, the instrument is deemed "GMP Critical".

Safety and Environmental Assessment

EHS instruments are assessed as per the criteria defined below. To be EHS critical, each instrument is critiqued against the following questions:

1. Would failure or inaccuracy of the instrument result in a major accident or environmental reportable non-compliance event?
2. Would instrument failure result in a possible unsafe situation or environmental control or escalation of consequences?
3. Does the instrument mitigate the consequences of an emergency or environmental event?

If the answer to any of the above questions is "Yes" and there are no other parallel or downstream components that would detect failure of the instrument, the instrument is deemed to be EHS Critical.

If responsible for safety instrumented systems on site, it is imperative to understand the specific responsibilities under the applicable safety standards. Adequately maintaining safety instrumented systems so they can function properly when required is a must. Through maintenance, inspection and testing procedures ensure safety systems maintenance is consistent. Process monitoring devices along with plant parameters and activities are critical for both operational reasons and in assisting with fault finding should a problem occur. It is vitally important that these devices are fitted on plant equipment in strategic parts of the process to guarantee optimum process monitoring at all times.

In an automated product mixing system, which may have very short cycle times in some of its processes, a company needs to be able to control and analyse product quality on-line and in real time.

Any delay in measurement can easily lead to excursions in concentrations of mixtures before being detected and ultimately may lead to inferior product being made which may have to be destroyed or reworked which directly leads to less profits off the company's bottom line. The key is to optimise the company's processes through increased access to **'more of'** and better relevant data by using an instruments advanced diagnostics and not just the primary measured value it gives out to a DCS system. It is vital during the design and installation of any system that the best and most accurate pieces of equipment are installed, the initial cost may be high but the benefits of virtually trouble free operation will far out way the initial equipment installation costs in the long run.

Always apply **'Zeroth's Law'** when dealing with temperature issues on any piece of equipment or process, it basically states: "If two thermodynamic systems are in thermal equilibrium with a third, they are also in thermal equilibrium with each other".

If two temperature probes (electrical or mechanical) are monitoring a piece of equipment or process and are both reading the same temperature, insert a third calibrated probe into the process to verify actual temperature, if all 3 probes are reading the same temperature, then they are in equilibrium with each other and the temperature readings can be deemed to be correct. Don't naturally assume that because two probes are reading the same temperature that the process is at this temperature especially if the process is 'temperature critical', always verify with a third calibrated probe.

Personnel must know the **thermodynamics** of the site's utility systems (Thermo meaning 'heat' and Dynamics meaning 'Power'). Thermodynamics studies the movement of energy, how energy instills movement and how that energy can be exchanged between physical systems as heat or work; it mainly involves changes in temperature, pressure and volume.

N.B. The importance of **fully functional strategically placed instrumentation devices** with redundancy monitoring/measuring pressures, temperatures, flows etc. cannot be stressed enough e.g. **a steam system.** It is vital to a plant's **safe and smooth operation** that these instruments are suitable for purpose, accurately calibrated and online at all times.

E.g. A typical example of a **heat exchanger** would be an electric kettle, where a 2 kW electric heating element in a kettle will bring 1.7 litres of water to a boiling point of 100°C in approximately 5 minutes. Electrical energy being converted to heat energy, thus boiling the water, other examples are immersion heaters, electric showers, gas central heating boilers etc.

Now multiply the above kettle kW rating of 2 kW's by 5,000 where you have an industrial facility which may have 2 or 3 large **'water converted to steam'** boilers which burn either gas or oil to superheat the water in the boiler to generate steam pressures of >8 Bar with temperatures of > +140 °C . This **steam** is then used to heat the various heat exchangers in strategic areas on site which supply heating via pumping systems using other liquid media such as water, glycol etc. to the entire plant i.e. central heating systems, air handling units, process vessels heating coils etc.

Steam heats these various heat exchangers circulating liquid to a set temperature as dictated by the control system via TT's and automatic control valves. Steam gives up its latent heat in the heat exchanger and is converted back to water condensate through a steam trap and back to the boiler for recycling. If an industrial boiler is capable of delivering 10,000 kW's of steam at a certain pressure, volume and temperature, the maximum heating load on site cannot exceed what is being delivered to it, i.e. if there are ten 1,000 kW's heat exchangers providing heat to processes which account for the 10,000 kW's of steam being delivered to plant (not allowing for inefficiencies & heat losses). More heating loads cannot be added (i.e. more heat exchangers), if another 1,000 kW heat exchanger is installed, it will be to the detriment of the other 10*1,000 kW heat exchangers already in place and will lead to decreased temperatures across processes.

Typical Steam Distribution System:

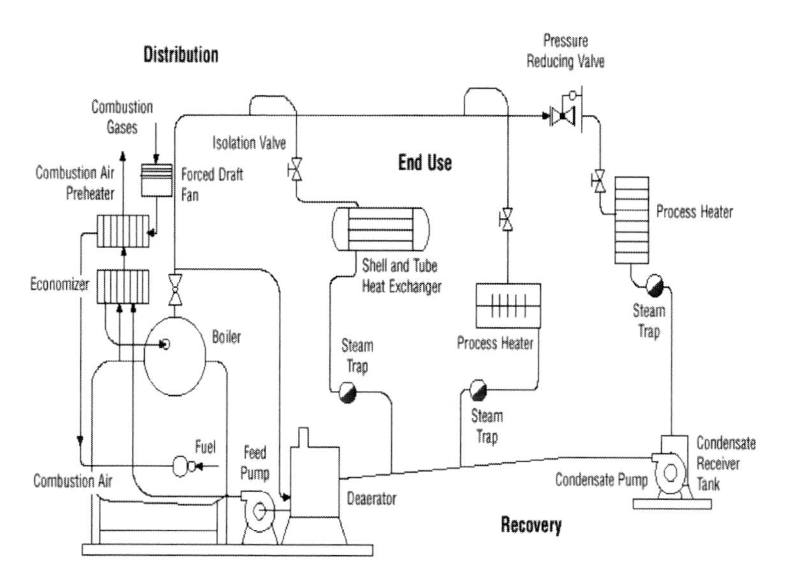

Safety Tip: Water hammer – Health & Safety Hazard

Waterhammer must never be tolerated, any system, which is properly designed and operated, will never experience this water hammer phenomenon, which is so often attributed to **'Steam Systems'**. Water is virtually incompressible. In a steam system, the normal steam flow is at probably some 35 m/s. If water (condensate) collects in the bottom of a badly aligned pipe (Fig.1), the steam velocity will cause ripples on the surface of the water. The steam will blow these ripples into waves until a wave is high enough to fill a pipe (Fig.2). There is then an incompressible liquid piston travelling along the pipe at steam speed. This whole pocket of water, picked up by the steam and carried forward in a solid column, is taken to some point down the line where there may be a change of direction or an obstruction such as a valve (Fig.3). The water is brought to a sudden halt at this point and the energy it contains, by virtue of its movement (kinetic energy), is suddenly converted into pressure energy. This sudden pressure causes large banging/rattling sounds and vibrations throughout the system that may cause considerable damage.

If the water hammer is due to the condensing of steam during distribution in steam pipes it can be eliminated. **Proper alignment** of the piping system ensures a continuous fall in the direction of flow and this must be combined with an adequate number of good drain points. It is particularly important to drain any low points in a system and to make sure that **steam traps** are working properly.

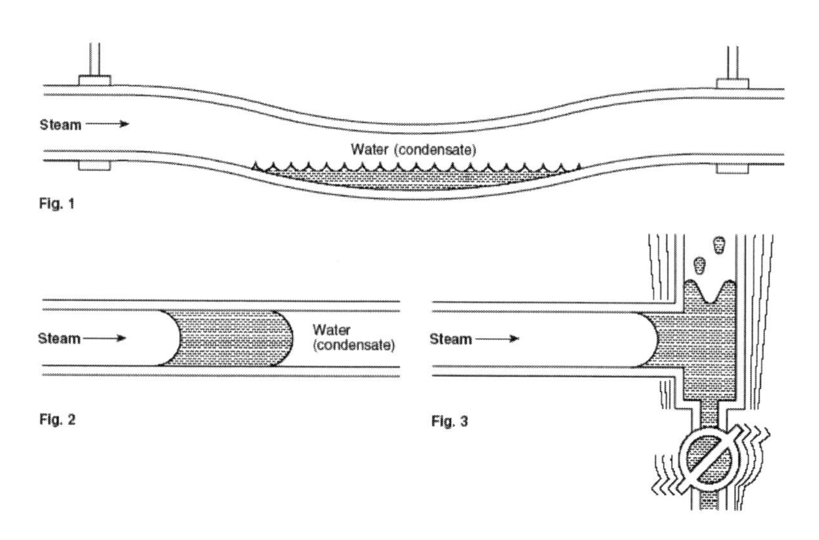

Fig.1 Poor alignment of a steam piping system:

N.B. Waterhammer can damage steam traps, valves, steam meters, pressure reducing valves, make joints leak, fracture pipes and **can cause serious injury to personnel.**

Steam pressure is very powerful and aggressive, when opening a wheel valve to pressurise a system, **do it very slowly.** Turn the wheel valve a quarter turn every 60 seconds or so or maybe longer, let pressure build up slowly.

N.B. The key to all of the above is to ensure that the steam system's entire network throughout a plant is properly installed. Personnel must know the plant's heating systems capabilities i.e. flows, temperatures, pressures, volumes, pressure reducing valves, steam traps and ensure it is capable of actually maintaining the heating needs on site.

Tip: Avoid 'the error of parallax', this may occur when reading any mechanical instrument from an angle, it is important that a person's head is held directly level with the point being measured to ensure an exact reading is being observed.

Instrument Calibration & Validation Overview

- Fully documented, properly indexed and stored calibration reports, calibration certificates with individual serial numbers for all relevant instruments and process instrumentation measurements are essential in ensuring the smooth running of a plant in a safe and cost effective manner.

- It is vital to have the onsite capabilities of being able to demonstrate that each component of the measuring system is functioning to specification.

- If responsible for safety instrumented systems, it is imperative to understand the specific responsibilities under the applicable safety standards.

- Instrument Validation means establishing documented evidence which provides a high degree of assurance that a specific instrument will consistently produce accurate readings, meeting its predetermined specifications and ensures that the specific instrument is working correctly and fit for purpose.

- Instrumentation devices that monitor process and plant parameters and activities are critical for both operational reasons and in assisting in fault finding should a problem occur.

- All calibration certificates must be traceable to the National Certifying body.

9

Trouble Shooting

"The process of diagnosing, locating and correcting malfunctions".

A typical industrial plant Compressed Air System:

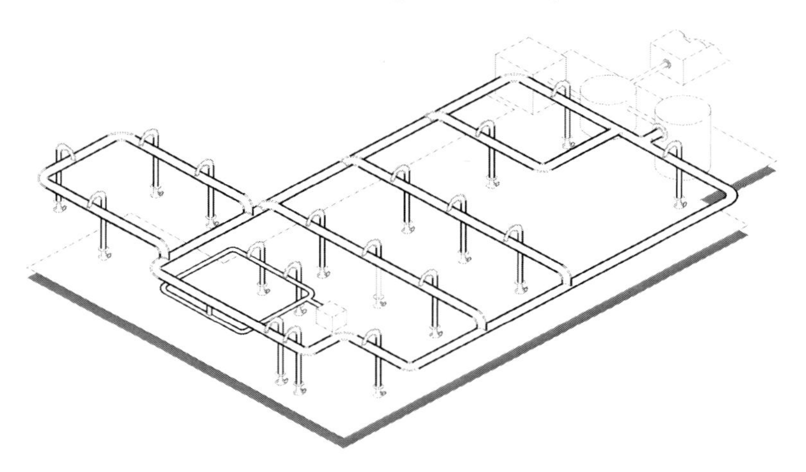

Trouble shooting is as much an art as a science. Disciplined problem-solving and long standing proven fault finding methods and techniques, teaches personnel never rush to a conclusion - examine the facts first, determine causality and judge accordingly, then determine a course of action. **Creative problem solving** derives from a person's ability to adopt multiple viewpoints and ask the right questions. It requires brain storming, solution seeking and proper implementation. It is the ability to identify and define problems as well as the generation of effective solutions.

Guided by theory, learn from experience.

For effective problem solving, regardless of how difficult a technical engineering problem may be, **personnel have to work through it** using a logical and systematic approach.

The problem must first be defined, cause identified and the correction properly made in a 'step by step' methodical manner. Having good problem solving skills is of real value because often personnel may have to work alone, isolated from their peers or manager. In any job, creativity and imaginative thinking are highly regarded. Even the most technical positions require the ability to think 'outside the box'. So never underestimate the power of innovative problem solving.

Try not to get caught in the trap of blindly following instructions without first understanding them. Without the understanding or 'know how' of where the source of a problem is, it will be difficult trying to resolve it. With little knowledge of how a piece of equipment operates, there will be little chance of fixing a fault, especially, if a complex one has developed. A company must always try to ensure that they have the right person in place, who has the right knowledge, relevant information, with the right tools (Hardware – hand tools or software - PC) and at the right time.

E.g. the same analogy applies if sitting for an exam, if you have done your proper preparation study, it is a great feeling sitting there, knowing you are well prepared. The same goes for a control system that you have to maintain.

N.B. The information structure that is in place on site will determine how easy or hard a job is going to be. There is no point in a company having all the relevant information, if it is very difficult to find.

Remember, knowledge is no load to bear. It is vital to understand a piece of equipments operation and capability and understand its associated written instructions clearly. Go through the instructions 'line by line' until it is fully understood what is written before carrying out any maintenance. Seek clarification of the instruction from the equipment manufacturer if need be. The key is to be able to manage the equipments' vulnerabilities and constraints and be able to verify its continued safe operational performance and how to recover from possible total failure.

Avoid 'knee jerk' problem solving, as much as team members would like to impress their peers or manager on their fault finding capabilities, they must not be tempted into taking chances. They must **keep their discipline**, stay focussed on proper fault finding techniques and adopt the right approach every time. **Ensure the process of solving the problem is managed.**

There are 5 items a person must have and understand to carry out proper methodical fault finding on a particular control system:

1. Actual Design Specifications. (Design Philosophy, **URS** - User Requirement Specification, **FDS** - Functional Design Specification, **SDS** - Software Design Specification)

2. O&M manuals (Principle of operation, Conditions of Start, Process Flow Diagrams, Block diagrams, Decision trees and flow charts etc.)

3. 'As Built' P&ID's (Process and Instrument Drawings).

4. 'As Built' electrical circuit diagram drawings.

5. 'As Built' mechanical drawings.

All of the above **original** documentation must be kept in a waterproof, flame retardant, controlled cabinet or area. Only signed, controlled, up to date copies of the original drawings and associated documentation can be allowed to be used by personnel on plant and must have the signature of the systems administrator who documents all activities surrounding these essential documents. Drawings issued by the Engineering department will be stamped 'For Information Use Only'; this stamp should have the signature of the issuer, the name of the recipient of the drawing and the date of issue. It is the responsibility of the holder of the drawings to ensure that the drawings held are the most recent revision. This is vital in ensuring any piece of equipment that has been modified is reflected in all documents via a change control procedure to ensure safety and compliance.

Change Control Procedure definition:

A change control procedure is a process of handling proposed alterations to systems that have been previously designated as fixed, whereby any planned change to a particular system is identified, evaluated, approved, implemented and documented. The purpose of change control itself is to ensure that change occurs in a controlled and safe manner and also to maintain the validation status of existing systems and to initiate a control process for new systems. It also gives all affected stakeholders the opportunity to review and control any subsequent changes to their systems and ultimately to 'sign off' before the change is carried out.

This may seem a total waste of time, however, if a change a person has made hasn't been documented, 6 months from now, one of their colleagues might have to fault find the same piece of equipment and realise a change has been made but they do not know the reason why. This may lead to hours of downtime, trying to decipher why the change was made in the first place. The initial person's actions could lead to an accident because other people, unaware of the changes made and the reasons why, caused an unexpected action when the undocumented change was put back to its original state.

There must always be a rational, logical engineering and operational reason to make a change on a system. This change must be approved by the affected departments' managers via change control and filed with all relevant documents pertaining to the change updated. Sometimes situations and circumstances dictate that all of the above cannot be in place before making the change. Personnel must always get their managers approval before any change is made and ensure it is documented retrospectively.

All team members must familiarise themselves with the plants equipment, be comfortable with individual pieces of equipment operations, know the **instrumentation design specifications, capabilities, sounds, smells, temperatures, pressures, flows etc.**

Personnel must gain a full understanding of the 'As Built' process, instrumentation, electrical and mechanical drawings. Physically go to the systems on the plant floor and 'walk down', the relevant drawings. Get to know where the control valves, hand valves, pipe work (pipes respective colour coding), individual electrical control panels and the associated pieces of equipment i.e. temperature transmitters, pressure transmitters etc. are positioned. Get to know why they are there, what purpose they serve and how they work. **The importance of the above cannot be stressed enough**. The team must dedicate time to this crucial familiarisation process whether they are on a large site covering a number of square kilometres with individual processing plants scattered throughout the area or a small site under one roof.

Proper, well maintained, visible and legible labelling of all equipment must be standard across any plant. **It is essential** not only for operational reasons but also when trying to resolve a fault which needs to be diagnosed quickly. The labelling should include control valves, hand valves, pipe work, electrical panels and the associated pieces of equipment. Even the individual cables/air lines and there associated connectors must be numbered and labelled according to the 'As built' drawings. If parts of equipment on plant are not labelled, personnel must make it their business to do so or get it done.

Heroic maintenance engineers, those who handle the exceptions and come up with all the answers, yet provide no accurate information of how issues were resolved and do not generate corrective actions (CAPA's) to prevent future reoccurences are a liability to a company.

CAPA: Corrective And Preventative Action: Defines the method by which Corrective and Preventive Actions are raised, monitored and closed out with the intention of preventing a reoccurrence and ensuring the improvement of a business's Operational, Safety and Environmental systems.

It is a very bad reflection on a person if having left the plant permanently; their successor cannot pick up where they left off relatively easily. There will always be a transition phase of learning. If the accurate, relevant information is already in place, this transition period should not be difficult, if one has done their job correctly. A person's reputation based on how well the systems they left after them were managed and maintained will follow, wherever they go.

Trouble Shooting Overview

- The process of diagnosing, locating and correcting malfunctions.
- Trouble shooting is as much an art as a science, creative problem solving derives from one's ability to adopt multiple viewpoints and ask the right questions.
- Problem solving requires brain storming, solution seeking and proper implementation.
- Having good problem solving skills is of real value because often personnel may have to work alone, isolated from their peers or manager.
- Try not to get caught in the trap of blindly following instructions without first understanding them.
- Personnel must keep their discipline, stay focussed on proper fault finding techniques and adopt the right approach every time.
- Team members must gain a full understanding of the 'As Built' process, instrumentation, electrical and mechanical drawings and physically go to the systems on the plant floor and 'walk down', the relevant drawings.

10

Fault Finding

Remember the old adage, *'Be prepared and always be proactive in failure prevention'.* Personnel must know the plant they work in, its respective systems, know what to do and how to adapt should a control system (hardware or software or both) start to shut down intermittently or fail completely. It is vital to be familiar with the underpinning fundamental physical principles and properties as are applied within a piece of equipment that is about to be worked on.

E.g.: *a pump is pumping liquid to a tank via a pipeline, an inductive flow switch is fitted into the pipeline to detect the liquid flow and to protect the pump against dry running, it is critical one understands the flow switch's'* ***'Principle of Operation'****, how it works, in other words:*

An inductive flow switch, normally fitted on a vertical pipe has a **ball bearing** fitted within the confines of the housing of the inductive switch, this ball bearing moves upwards/downwards when flow is established/not established i.e. the ball bearing moves in or out of the induced field of the electrical flow switch telling the control system (PLC etc.) it has detected flow/no flow of liquid in the pipe. *This may seem like common sense to an experienced fault finder, to an inexperienced one, it may be very difficult to decipher.*

A person may have studied all the background literature and fault finding methods about a piece of equipment but sometimes it is possible that two normally obvious problems combine to provide a set of symptoms that can be totally misleading and send the person completely in the wrong fault finding direction.

Being aware of this possibility and avoid solving the 'wrong problem' is what must be recognised and understood.

There are many different types of faults, from minor to major, ensure any action that is carried out does not exacerbate the fault. Can the piece of equipment run safely, even if not at 100 % effiency until personnel feel confident in carrying out the repair?

Don't turn a minor fault into a major issue by one's actions.

Plan, Execute, Test and Document.

N.B. 'Plan the work and work the plan'

The extent of preparation should be directly proportional to the size of the problem to be rectified, i.e. replacing a light bulb versus replacing a piston in a 10,000 HP engine.

Tip: Personnel must remind themselves to make a **conscious decision** with regard to intended actions and how they are going to go about them. Replay proposed actions in their minds and plan strategies regardless of the type of work that is to be carried out. Be prepared for the inevitable obstacles as they arise. Try to avoid, where possible, making big decisions regarding the dismantling of a machine if under stress, physically or mentally tired, suffering from sleep deprivation, feeling unwell or in a hurry. If this situation arises, take a 'time out', think about what has to be done and the possible consequences (if any). Decisions made under these conditions can prove very costly to both personnel and equipment if a mistake is made.

Make sure decisions are based on accurate information.

E.g. ***a fault has occurred in a machine:***

What are the **'Conditions of Start'?** *Personnel must not* start changing parameters (P.I.D settings) in controllers, pushing buttons, turning knobs, closing/opening hand valves etc. if they are not positive they know what they are doing.

Discipline must be maintained at all times, no matter how much a person would like to press a button, open a valve etc. They must **keep their hands to themselves** and ensure any action taken is based on competent engineering fact. One must try not to suffer from **'Analysis Paralysis'** either (**thinking gets in the way of doing**), where they may be receiving multiple pieces of information from all quarters, possibly leading to 'information overload', using their 'process of elimination skills' here will be vital. **Do not** make whole scale changes, take a structured approach and take one step at a time. 10 -15 minutes of thought can save hours of downtime and personal anguish.

E.g. replacing a faulty PCB (Printed Circuit Board) in a machine:
Before discarding the faulty unit, ensure to observe any 'jumper connection links' or any 'switch set up arrangement' that may be in a bank of switches on the faulty PCB itself and replicate exactly both links and settings on the new PCB prior to fitting.

Personnel must use their natural senses:

- Look out for anything unusual, oil stains, leaks, burn marks etc.
- Use their nose to smell anything out of the ordinary, burning oil, rubber etc.
- Get to know the noises a piece of equipment generates, check for any excessive vibrations by actual touch, this could be first sign of wear and tear of a bearing.
- Listen out for anything that doesn't sound right.

Predictive Maintenance: (CM) Condition Monitoring or a (CbM) Condition based Monitoring programme has many benefits including overall equipment effectiveness (OEE), reduced mean time to repair (MTTR); increase Return on Net Assets (RONA); reduced unplanned and planned downtime duration and lower inventory cost.

A company must constantly try to improve 'early asset failure recognition' to maximise equipment reliability.

3 examples of condition monitoring services are:

1. **Vibration Analysis:** measures change in vibration intensity on mechanical equipment when machine condition begins to degrade. Vibration analysis is probably the best known CbM tool. It detects equipment vibration levels affected by factors such as misalignments, unbalance, displacement, velocity, looseness, eccentricity, defective bearings, resonance, electrical problems and aerodynamic/hydraulic forces. Equipment misalignment alone results in wasted energy and causes premature equipment breakdown. The best way to reduce maintenance and operating costs of rotating machinery is to align the machinery correctly in the first place. The goal is to identify changes in the condition of a machine that will indicate a potential problem.

2. **Oil analysis:** detects contamination or degradation of oil which indicates machine wear. Correct oil lubrication is critical for the efficient operation of equipment and must be maintained according to the manufacturers' instructions.

3. **Infrared Thermography:** detects variations of temperatures in electrical, mechanical, infrastructure and process equipment. Thermal Imaging is an important part of any preventative maintenance program. Infra red thermal

imaging inspections will benefit a company by providing real time data e.g. temperature readings, to determine defects that can lead to equipment failure e.g. motor running too hot. Applications include infra red testing and inspection of electrical distribution systems, motors, heat exchangers etc. helping to diagnose abnormal 'hot spots' in pieces of equipment e.g. M.C.B. failure (poor electrical connections), damaged motor bearing, or 'cold spots' partially blocked heat exchanger leading to inefficient operation. An infrared camera can be used for a broad range of predictive maintenance applications enabling problem areas to be clearly seen on a crisp thermal image.

Never underestimate the difficulties of any repairs to be carried out.

Always exercise caution and systematic thought, use logical and deductive reasoning at all times. If a person makes a decision to attempt to carry out the work needed to investigate or actually remedy a fault, they must ask themselves:

- Am I competent? Establish what they know and what they don't know.
- Have I got enough knowledge, technical know-how, experience, technical manuals/ information to hand and level of expertise to carry out such a repair?
- Have I the specialist equipment and tools needed to carry out the work? Machines will normally have complex pieces of individual components attached to them e.g. a gearbox.
- Have I got all the proper spare parts available to me?
- Is the vendor/manufacturer who supplied the equipment readily available? It is vital to have the emergency mobile and landline phone numbers of the vendors and manufacturers service people readily available for expert advice and help should anything go wrong.

- Am I capable of doing the repair and have I the right tools and equipment (e.g. hydraulic lifting jacks, portable power tools) to carry out such a repair?

N.B. When using tools and equipment, ensure they are safe to use and fit for purpose. Ensure they can be accounted for and know where they are at all times, the reasons being:

1. If tools are left loosely on a pipe or steel girder, they can fall and hit someone, causing serious injury.
2. Trying to find for example, a wrench that has fallen into a myriad of hydraulic pipes, hoses or cables can be very difficult and time consuming.
3. If a screwdriver falls into a rotating mechanism accidentally and is not noticed, there could be serious consequences should the machine be turned on, the mechanism could be jammed or damaged. The screwdriver can also become a missile if ejected from the mechanism at a fast speed which could cause serious injury.
4. A 10 mm steel hand spanner that has been mislaid and fallen into electrical switchgear unnoticed, could have disastrous consequences if the electrical equipment was switched on, possibly causing short circuits across 'live' contacts resulting in a large flash/electric arc which could cause serious burns to a person or be possibly fatal and may also cause severe damage to the electrical equipment itself or cause a fire.
5. **Portable equipment,** by its nature, is more susceptible to damage than fixed electrical equipment. It is also more likely to be used in different environments and is often directly in contact with the user. If using this type of equipment (ensure it is PAT tested – Portable Appliance Tested). Ensure the AC transformers, power leads and power tools and associated plug tops are all in good condition and fit for purpose before using them e.g. RCD protected, properly fused, no cuts, **no damaged or non - standard joints** including taped joints are evident in the power leads.

Ensure also that power tools themselves (drills, grinders etc.) whether they are DC or AC powered have no sign of physical damage to the outer cover of the equipment or obvious loose parts or screws. Ensure the outer covering (sheath) of the cable is being properly gripped where it enters the plug or equipment and no coloured individual wires are visible.

6. When using long power cables it is important to take into account the amount of **voltage drop** that can occur, as it can cause the cable to become hot and unsafe i.e. As the current passes through a longer and longer conductor, more and more of the voltage is "lost" (unavailable to the load), due to the voltage drop developed across the resistance of the conductor.
7. Ensure any lifting equipment (hydraulic jacks, winches, block and tackle etc.) used, is properly serviced, certified, has no apparent physical damage present and is fit for purpose.
8. Tools and equipment can be very expensive to replace should they be abused, mishandled, mislaid or stolen.

Abnormal Operation Analysis and Correction:

Four logical steps are required to effectively analyse an operational problem and make the necessary corrections:

1. Define the problem and its limits.
2. Identify all possible causes.
3. Test each cause until the source of the problem is found.
4. Try to establish what was the 'initiating factor'? i.e. what caused the fault in the first place and make the necessary corrections.

N.B. Ensure any work carried out does not cause loss of control or escalation of consequences in the event of equipment failure.

The First Step: is to define the limits of the problem:

- When a problem develops, ensure a good problem definition is created; compare all information with normal conditions. Up to date P&ID's, logbooks or PDA's with daily recordings of key parameters, defining expected acceptable limits of the equipment's operation will be invaluable here. Preventative Maintenance Records and Work Instructions are also a good source of information on a piece of equipment's normal operating mode. If well maintained, these can be a very useful when solving an intermittent problem/fault.

- **'Resource Management'** plays a key role when trying to resolve a problem. This is where people skills will be vital. The best option is to speak to the person who reported the problem, if available. The information gleaned will be invaluable when fault finding, especially, if the person is very experienced in the equipment's operation.

- **Do not be afraid to track down the operators, other maintenance personnel or instrument technicians for input in resolving the problem. Two heads are better than one. Ask questions to help gain clarity on the details of the problem. Communicate with others as clearly as possible to avoid misunderstandings. Do not dismiss a person's comments or observations out of hand, respect the opinions of others, it may have no relevance or it may be the exact piece of information being looking for.**

- Knowledge, experience, up to date O&M manuals, engineering drawings and consistent records are the basis for avoiding the unusual especially when applying **RCA (Root Cause Analysis)** techniques. When the 'direct cause' of a problem is immediately evident e.g. of 1,000 machine parts being made per hour on a piece of equipment, 100 of them are undersized and are rejected. Use of the '5

Why's' principle can now be implemented i.e. personnel may have to ask 'why' 5 times before getting to the 'root cause' of the problem. As well as looking for 'root cause', they will also have to take into consideration the 'contributing causes'. Probably they would not have caused the problem on their own, but, increased the likelihood of the problem occurring. Lack of 'follow through' on RCA lessons is a common weakness and is often due to an ineffective logging or a progress tracking system.

- Hearsay, opinions, beliefs or anecdotal information based on or consisting of reports or observations of unskilled observers must be received with caution as these observations and reports can lead the fault finder in the wrong direction. Ask if they are basing their observations on gut feeling or operational, process and engineering fact? It is vital the latter takes precedence when considering the best way to tackle the problem. Remember the old adage 'Facts not opinions'.

- Gut feeling is one thing, but causing **€100,000 worth of damage** to a piece of equipment because of 'gut feeling' fault finding techniques is another and then trying to justify what was done which caused the damage and made matters worse. If personnel work a shift pattern and come across day to day engineering issues that need addressing, they must not leave it to others; they themselves must pass on all relevant information to their direct supervisor and to the oncoming shift. Include all relevant personnel who may be directly affected by these issues. Inform verbally, by written shift report or on email (if it exists on site). **The most important point is the communication and follow up,** even if this means phoning the plant after personnel have gone home to ensure the relevant information has been received and understood by the on site shift personnel. Never underestimate how essential this follow up is.

Simple misunderstandings can lead to huge and costly delays.

- Ensure whatever message is passed on is clear, unambiguous and fully understood, whether to internal shift personnel, other departments or vendors. Do not leave an email, verbal or written message open to interpretation. It may seem obvious to personnel, because, they may be familiar with the system, to others it may be very difficult to decipher. In other words 'Add the detail'. This will make everyone's jobs easier. **N.B.** Just because a message has been passed on, doesn't mean it is no longer the onsite shift's problem and it automatically becomes the incoming shift's problem.

- When a message is given verbally, ensure it is repeated it to the person, so that they understand how it has been interpreted. A person must have their mind in a 'thinking gear'; removing any other distractions that they may be thinking about. Concentrate on what the person is saying, write it down if need be, if something doesn't sound right or doesn't make sense, make sure it is clarified with the person delivering the message. Be exact in all communication and correspondence to all. Leave no room for guesswork. Always act in a professional manner. Ensure to follow up with the other members of the team on the successful resolution to a problem, regardless of who resolved the issue. **Close the loop.** It is vital to hand over the plant in a safe, clean and fully operational state to the incoming shift. Go the extra yard to ensure this happens. Personnel must leave the plant the way they would like it to be left when **they** are coming on shift.

- **Tip:** During and at the end of a working day, take 5- 10 vital minutes to recap what work was carried out. Has the plant been left in a fully operational and safe state? i.e. Did personnel:

1. Leave anything running which should now be switched off?

2. Take any system or individual automatic valve/motor out of 'automatic mode' to facilitate maintenance and did not put the piece of equipment back into 'automatic mode' and 'test it' once the work was completed?

3. Close or open a hand/automatic valve and didn't put it back to its original setting or position?

4. Remove the 'lock and tag' from an isolator?

5. Checked and double checked their work before they 'Signed off' on the permit to work?

- A small note pad should always be kept and little reminders written in during the day to jog a persons' memory, at the end of a shift, of what actions were carried out and what was isolated/turned on so they could carry out their duties. They should then go through the actions 'one by one' and to guarantee 'peace of mind', ensure everything that was worked on is back on line and 'fit for purpose' before leaving site.

- If a person doesn't know how to fix a fault, they must say so and '**Not bluff it**', then ask for more time on the issue so an informed factual decision can be made. They must realise and recognise the limits of their own technical capabilities. If they've been working on a piece of equipment for hours and cannot resolve the fault, they must '**Stop**' and take stock of the situation. Don't waste anymore time. Recognise they have exhausted their fault finding capabilities and need help. Use the time instead to escalate the issue. Inform the relevant people or department(s) of their predicament, then start looking for advice from other colleagues or vendors. Research more information and ultimately learn from the experience.

- **'Never Assume'**, when the word 'assume' is broken up it can be interpreted as making an 'ass' out of 'u' and 'me'. **Don't fall into the trap of making assumptions.** Assumptions most of the time lead to incorrect decisions and can prove very costly. Always approach the fault with the actual manufacturers' operational manual of how the piece of equipment 'actually works', **not** the anecdotal information of 'this is how we always operated it'. The O&M manual is an essential part of the machine and must accompany it during its entire lifetime. Just because a piece of equipment has been operated in a certain way over a period of time, doesn't mean that this is the proper 'mode of operation'. There may have been an undocumented modification carried out. Reference to the manufacturers' manual, will help identify any modifications done to the original design.

- **Maintenance procedures are put in place to ensure any work carried out, is done according to manufacturer's instructions. In that way, repeatability and accountability is maintained at all times.** If there are **not** proper Preventative Maintenance schedules in place over the lifecycle of a piece of equipment, complacency can set in. Safety interlocks might **not** be working as they should and a 'work around' may have been put in place. The piece of equipment could then be operated in this incorrect manner, with the potential for an accident or incident to take place. The piece of equipment must be operated according to the manufacturers instructions, no 'work arounds' should ever be put in place, 'Moving the goal posts' to get a result is not an option.

- **The key to minimising equipment problems are scheduled service intervals and routine inspection.** Written records indicating date, items inspected/replaced, service performed and machine condition are important to an effective routine maintenance program. From such records, specific problems can be identified and solved routinely. To avoid breakdowns, a piece of equipment may have operational parameters within which the machine can work safely i.e. normal running temperature of 70°C, with a +10°C safety margin.

- If a piece of equipment which has been operating safely and properly according to the manufacturers instruction over a number of years suddenly stops for no obvious reason, personnel must proceed to make a list of all deviations from normal operating conditions.

- ***Do not rush in and make wild guesses.*** Think logically before tackling the issue head on. **Take time** to make sure the proper documentation is at hand to deal with the problem. 10 - 15 minutes of preparatory work to refresh a person's memory on how a system operates can save valuable time as they try to find the fault and also, may be less likely to make mistakes, thus diagnosing the fault more quickly. Remove any items not relating to the symptom and separately list those items that might. Use this list as a guide to further investigate the problem.

The Second Step:

Is to decide which items on the list are possible causes and which items are additional symptoms. Again **the extent of preparation should be directly proportional to the size of the problem to be rectified.** Use the systematic approach i.e. has a decision tree, fish bone diagram etc. been developed based on the piece of equipments' operation and real time recordings of relevant parameters? Using these types of diagrams and actually writing down all the possible conditions and potential problems regarding the fault with other members of the team can help build up a picture or pattern that can be used to assist in finding the fault more quickly.

This type of 'brain storming' is only as good as the people who are carrying it out. It is vital to have the right mix of expertise who have a good understanding of the actual problem and the affected equipment. The team should comprise of both engineers and process personnel. The team should involve any other department who they think may be of benefit in resolving the problem.

The Third Step:

- **Do not** alter several things at once as it may never be known what caused the problem, if the problem rectifies itself.

- Identify the most likely cause and take action to correct the problem. 'Think' before any action is taken. **Do not** change a setting or a parameter on a piece of equipment if not sure of the ramifications of such a change. **Do not** change a setting or parameter unless it can be returned to its original position.

- **In solving a wrong problem, a new problem can be created.**

- **Do not** withdraw a piece of equipment from a unit unless it can be put back to its original state. Observe the orientation of how each component is interlinked with other components. Be extremely careful when removing multiple individual pieces of equipment to get at a faulty part, as any breakages can prove very costly and increase downtime dramatically. The importance of making sure to retain every nut, bolt, anti vibration washer, screw etc. in a safe place during disassembly cannot be stressed enough. Lay them out neatly in a structured sequence on a bench and ensure they all can be accounted for.

Tip: Get a piece of cardboard and some masking tape, as you remove the screws, nuts and bolts etc. tape them to the cardboard in the sequence you took them out and replace in the reverse sequence after repair is complete. If after reassembly there are some bolts and nuts left over, it will be clear, reassembly hasn't been successful.

- Trying to find or replace a special length bolt, screw etc. that has gone missing can be very time consuming and frustrating. Also take into account that some bolts may be longer than others for a specific purpose and must be replaced exactly as they were removed.

Tip: Get another piece of cardboard and use it as a template. Draw a rough sketch of the part that is being dismantled and as the bolts are being taken out, push them through the cardboard as a reminder how they should be replaced. Digital cameras or mobile camera phones can prove very useful in having actual real time pictures of the original installation to assist when reassembling.

- If a digital camera/mobile picture phone is not available, always have a **black permanent marker** to mark the pieces of equipment that are being removed.

E.g.: a mechanical seal on an agitator shaft has to be replaced. Draw a straight line through all the individual sections before disassembly and use the line as a reference when reassembling. Parts can then be realigned in their original position after carrying out the repairs on the faulty part.

- When reassembly has finished, mark each bolt, nut, screw with a little yellow paint/permanent marker after they been tightened or torqued to a manufacturer's recommended specification. By doing this, there is the added assurance that it can be visibly seen if any bolt etc. without a yellow mark has not been properly secured.

- If a fault has been misdiagnosed and the piece of equipment that was thought to be the cause of the fault turned out to be functioning properly, replace the original part, as electrical and mechanical parts can be very expensive and sometimes very hard to replace.

- If the symptoms are not relieved, move to the next item on the list and repeat the procedure until the cause of the problem has been identified.

- Take digital pictures of a smoothly operating piece of equipment i.e. (all mechanical/electrical settings and digital faceplates/screenshots). Photographs make great reference material should a problem occur. If a setting was changed accidentally or otherwise, it can be checked against the photo. Make sure they are stored in a place that is secure and easily accessible.

Tip: When removing badly rusted bolts, nuts, bleed fittings, be patient, use plenty of good quality penetrating spray and leave to 'soak in' (possibly hours) before loosening, reapply as necessary. Stripping threads or shearing off a nut, bolt or bleed fitting while trying to remove it, can prove to be very expensive and difficult to repair or replace.

The Fourth Step

- Once the cause has been identified and confirmed, analyse the fault or faulty equipment. Investigate why it failed. Remove the cause of the failure and put in a CAPA to prevent the fault reoccurring. Proceed to make the necessary corrections. Keep an open mind when implementing a CAPA. Formulate new questions to provide a different direction if need be, **it is insanity to do things the same way and expect a different result.**

- Test the equipment to ensure it is working properly and according to the manufacturers' instructions, specifications and guidelines before putting it back into service.

- **Keep incident reports,** learn from them and decide how the incidents can be prevented in the future. If a fault rectifies itself, always remain suspicious and carry out as many investigations as possible to determine what caused the fault in the first place. Most faults that self – rectify, normally fail when the piece of equipment is needed the most.

- **Be aware**, an undetermined cause to a problem that has rectified itself, leaves a situation that can happen again as no preventative action can be put in place to mitigate against the problem reoccurring.

There are 3 areas that must be clarified when dealing with total failure of equipment:

1. **Facts** – Deal with the actual facts that are evident. They must be written down and clarified. Focus and rely on facts and accurate data before dealing with anecdotal information.

2. **Conditions** – Have any environmental conditions changed dramatically? Has an unexpected instance occurred? E.g. lightning strike, excessive ambient temperature, freezing temperatures etc.

3. **Circumstances** – Has something happened or has someone done something to the piece of equipment that would not be normal and may have been missed during the fact finding mission? i.e. a change of a parameter or mechanical setting could have been made during the previous days' preventative maintenance schedule. Some faults may take months to manifest before failure occurs, e.g.

> **'metal fatigue'** on a structure may start with a single tiny crack. When mechanical stresses are physically put on the structure, the crack may get bigger and longer and may ultimately lead to 'breaking point' and catastrophic failure.

It is much easier to establish facts when the equipment is running, even if not properly i.e. noisy, excessive vibrations and intermittent stoppages. When the piece of equipment has completely shut down and with no indication of what is wrong (internal failure), the problem becomes a lot more difficult to diagnose e.g. motor seizure in a screw air compressor i.e. it could be a motor winding failure, motor mechanical bearing failure, mechanical screw failure etc.

What a person learns from one control system (hardware, software, process and instrument drawings etc.) can help dramatically when called onto another similar but not identical control system where a problem has occurred. This is called **generic learning** (i.e. what is learned on one system is brought to another).

A person must write good reports on how and what they did to remedy a fault. File them in a manner which can assist them or others if the same problem arises in the future. These notes can prove invaluable in saving time and effort should this fault reoccur. **If the work that's done is not documented, dated and signed for, it didn't happen.** There must be a record, whether it is a hand written document or electronically recorded of all activities carried out on a piece of equipment.

If a control system is running perfectly, it is vital personnel take the time to either **write down or electronically record** all parameters i.e. P.I.D. (Proportional, Integral, Derivative) control instrument loops settings, temperatures, pressures, flows etc. for all associated equipment within the control system. Keep these parameter lists in a safe fire proof place (not in the same building) where they can be retrieved should a control system crash occur and does not recover properly. This recorded information can be used as a checklist for the smooth restart of the system.

If a person has little or no experience on a particular control system, remember, that most control systems do virtually the same thing. The variation of control systems and equipment and how they are arranged to carry out a specific task safely is what they have to figure out.

Safety Management of control systems must be employed to ensure safe operation of a machine i.e. the control system has to be integrated with safety systems which are installed in the field in order to partially or completely isolate/shutdown equipment in case of maintenance or extraordinary operations. Guards and interlocks are designed and fitted in the interests of safety and UNDER NO CIRCUMSTANCES should the equipment be operated with guards removed or interlock switches overridden.

Control and safety interlocks are put in place to ensure proper safe operation of equipment; upon detection of abnormal operating conditions, a safety circuit or hardwired interlock uses mechanical means to place equipment and process in their defined fail safe state, thus ensuring the safety of personnel, plant equipment and environment. When fault finding, a person's job is to find out what the interlocks are and in what sequence they operate. If one can gain some understanding of the 'principle of operation', it will be easier to solve the problem.

Most machines built today have PLC's (Programmable Logic Controllers) fitted to them to operate and monitor the control system. Normally hard wire safety interlocks which work independently of the PLC are also part of the entire control system. If it has a HMI (Human Machine Interface) attached, personnel must try to navigate its matrix and find its graphical alarm screen for text that may tell them what the problem is. Most HMI's have actual graphics on the screens of the machine itself and the piece of faulty equipment will flash 'Red' if it has failed to start. If a malfunction is suspected but there is no diagnostic message on the HMI display, the 'conventional way' of fault finding will have to be implemented and physically go inside the equipments main control panel. Use the standard procedures described next as a guide to verify that the equipment hardware and process connections are in good working order. Always approach the most likely and easiest-to-check conditions first.

Before tackling the entire control system, try not to overlook the obvious:

1. **Electrical power** - is the correct mains voltage present? e.g. 400 VAC @ 50Hz? Be aware of back voltages especially if there are step down transformers and power supplies providing specific voltages (110 Vac, 24Vac, 24Vdc, 12Vdc etc) needed for the HMI's, control panels, instrumentation etc. These can give misleading back feed voltages and cause confusion. Always establish correct voltages are present starting with the machine's main electrical isolator and then methodically isolate each individual breaker or control fuse and all the while establishing that the correct respective voltages are present according to the manufacturer's specifications.
2. **Compressed air** - is the correct main air pressure present? e.g. 7 bar and where PRV's (pressure reducing valves) are used, are they set at there respective correct pressures? i.e. 1.4 bar, 3 bar.
3. Other utility supplies are available and online.

If it's a piece of equipment which uses a variety of utility systems such as compressed air, refrigerant cooling systems and Nitrogen for example, check that all are present and at the correct pressures required by the equipment. Pressure and flow switches are normally fitted to these utility services to ensure smooth operation, if not energised (fail safe) at all times, the machine will stop and in doing so, protects both personnel and the machine itself.

Check the safety interlocks:

- Emergency stop buttons are not pressed.
- All electrical isolators to internal motors etc. are switched on.
- Key switches are properly set (Hand, Off, Automatic).
- Micro switches on doors, guards etc. are working and engaged.
- Check for any physical damage done to the machines ancillary equipment (solenoids, automatic valves etc.). Check electric cables for damage, hydraulic and pneumatic hoses for leak or rupture.

The electrical isolator for the machine will be mechanically interlocked with the main electrical panel door i.e. the panel will have to be electrically isolated before the panel door will open. Once inside the control panel, check for any electrical literature and drawings that may be of assistance in finding out how the machine operates. Take the time to read and understand any drawing related to the machine that is available. The next step is to check for tripped circuit breakers, blown fuses, controller failure (i.e. no LCD display), any sign of burned contacts and investigate the cause of failure. If, on the other hand, there is no literature or electrical drawings available i.e. they have been taken from the control panel and not replaced, try to locate them immediately, if this is not possible, get a pencil and paper and start making notes and drawings. Try to start piecing all the bits of information together to gain a good understanding of the machine's operation.

Safety tip: When tracing cables in an electrical panel, ensure to wear electrically resistant insulating gloves. Watch out for 'nicked, cut or frayed cables'. Repair or replace immediately.

Most MCC electrical panels are now being built with 'IT intelligence' i.e. each motor in the plant will have its own cubicle 'plug and play'. It can be completely withdrawn from the main electrical panel and worked on. This panel has all the associated control equipment needed to operate the motor i.e. power on/stopped/running/tripped status, number of starts/ running hours. Conditions of the motor itself are monitored by a PLC system with a HMI attached to it. Reports can then be generated, disseminated and used to focus maintenance resources in the areas that most need them.

This type of automation infrastructure exhibits much more intelligence than conventional systems and greatly assists fault finding. Diagnostic functionality is improved because programmable systems offer constant test outputs with full diagnosis through the software – something that can't be done with conventional systems. The result is that faults are easier to identify and put right, so that downtime is reduced, as is the input from potentially expensive specialist engineers.

It is vital personnel familiarise themselves with the internals of the motor cubicle and its associated electrical drawing. There could be anything up to 10 individual main components involved. It is critical to identify each component and that they tally with the electrical drawing. It is also critical to actually **understand their respective functions and all the interconnecting 'component' interlocks** i.e. electrical isolator, MCB's, control fuses, contactors, digital timers, relays, current monitoring devices, overload protection devices, device cards to interface with the PLC/HMI system (which will monitor all these components 'in real time for tripped/failure status) + all associated cabling and connectors. If any one of these components fail, is there another readily available?

N.B. One of the advantages of having a HMI screen is that if all the 'ready to start' conditions/interlocks are not met before starting a piece of equipment (e.g. Overload tripped) the system will not make the 'Start command' available to personnel. The HMI alarm screen can then be interrogated to find out immediately what individual **'component'** has actually tripped. Investigation into the root cause of 'overload trip' can start immediately.

If after a period of time, it is believed one has figured out the machines 'Principle of Operation' and has discovered the fault, they must proceed with caution always being vigilant. The following would be a typical example:

E.g. *A 24Vdc electro pneumatic 5/2 way spring return solenoid valve which when energised/de-energised pushes and retracts a pneumatically operated piston has failed to operate and the valve remains non- operational, the following procedure should now occur:*

1. Check correct air supply to valve is present (check local pressure gauge, if fitted).
2. Check that the correct voltage to the coil of solenoid is present.
3. Check valve mechanism (listen for clicking sound, when solenoid coil is energised).
4. No clicking (either solenoid coil is faulty or mechanical actuator in pneumatic valve is mechanically jammed).

It is discovered that the 24Vdc coil of the solenoid which activates the valve's mechanical mechanism is faulty and **a replacement does not exist onsite.** This is where real maintenance kicks in. If a full, up to date inventory of all equipment on site is kept, personnel may be able to:

'Adapt, Improvise, Overcome'

Are there other machines like it on plant which are not as critical or not in operation, from which you can borrow a replica unit until a replacement is available? If not, can another type of solenoid be installed? Perhaps not the same vendor model but operates exactly as the original solenoid does. A good engineering practitioner's job is to provide safe options and keep the plant running. Unfortunately, most companies are focussed on cost cutting and to have expensive parts sitting on a shelf 'Just in Case' sometimes is not an option.

Keep an open mind. Try to grade (Scale of 1- 10) how important a piece of equipment is and the consequences to the plant, if it fails. Try to figure out all possible failure scenarios and try to put in place, measures to mitigate against such scenarios occurring. **This is called the 'What if' side of engineering.**

What might be mind boggling to some, may be blatantly obvious to another. Despite all efforts, sometimes it is easy to get caught up in a fault and get nowhere fast. **Don't ever be afraid, ashamed or embarrassed to ask for help.** Discuss the fault with other peers/manager. Get their input if they have also been trained on the piece of equipment, **good counsel can be invaluable.**

Call in expert help if available and learn from them, try to have in place, if possible, a direct contact with the vendor manufacturer if a problem occurs. The problem being encountering may never have happened before and was not foreseen by the manufacturer who designed the equipment. **There may be no magic answer.** Knowledge and experience of the equipment will play a vital role in diagnosing the problem, with the manufacturer's assistance.

N.B. If a new piece of equipment comes on site and is to be commissioned by the vendor's technical engineering personnel, ensure site personnel are with them every step of the way, no matter how busy they are. The time dedicated to this learning is invaluable. **The experience gained from the experts sometimes can not be found in O&M manuals**. The same goes for existing pieces of equipment on site that may need to be given an overhaul by the vendor. Ensure as many relevant key personnel on site are trained. Insist that personnel and time be made available to gain this knowledge and experience.

A person may work on a piece of equipment with either a low or astronomical monetary value. It may be a 'Space Heater' or a 'Space Ship'. The same fault finding techniques should be employed.

E.g. *an electric motor which turns a mechanical pump and pumps liquid into a tank starts to fail intermittently:*

The motors' overload, is a device which monitors the motor's maximum electrical current draw and is designed to ensure the motor itself stays within its design safety limits, starts tripping intermittently, stopping the motor.

NB. Under no circumstances should an overload be adjusted above the current (amp) rating on the nameplate of the motor.

Another problem then develops, the motor's current circuit breaker starts to trip. The problem is believed to be a motor problem. Despite standard electrical checks (proper voltage present on all phases, integrity of cable connections checks, winding continuity and short circuit tests) indicating no issues with the motors windings, the motor is replaced and the same problem reoccurs. **Considerable downtime has now elapsed, the problem still exists and the fault finder is still at square one.**

If a good installation qualification (IQ) has been carried out with all the original parameters and recordings of the actual motors' performance is available (benchmark or baseline figures), use this as a starting point. If over a period of time the motor current or temperature rises inexplicably, investigate immediately before the motor fails completely. If no information is available, read the maximum current rating on the motors nameplate and measure the running current. If the motor is running close to or at its rated current, investigate immediately. If it is a new installation, questions must be asked e.g. is the motor rated for the mechanical load? Can the mechanical load on the motor be reduced until the problem is found and rectified? Is the motor itself suitable for the environment it was selected to operate in? E.g. poorly ventilated area or excessive environmental temperature conditions.

The actual problem is found to be that the pump that is being driven by the motor is mechanically overloaded due to an ingress of solids (making it harder to pump) thus putting a larger load on the motor causing it to trip out.

E.g. *An electric motor which drives a large industrial air intake fan via belts and pulleys in an air handling unit (AHU) experiences high running currents and temperatures. The air flow from the circulating fan via ductwork maintains an air pressure between two areas in a process plant. This pressure is deemed critical and must be maintained at all times during processing:*

Check:
(i) Electrical current load against maximum load current rated on the name plate of the motor (it is vital an 'operating log' of the motor is kept and maintained). This log can help in determining what constitutes normal versus abnormal operations.

(ii) If the motor is running close to or at its maximum rated electrical load current capacity or over its maximum permissible running temperature, check:

What protection relays are monitoring the motor's operation and if activated on a trip condition, will stop the motor:

- A voltage monitoring relay which guards the motor against the damaging effects of phase loss, under and over voltage, phase imbalance, phase reversal and voltage quality of incoming power line.

- A current monitoring relay which provides the motor with under and over current detection.
- The motors thermistor or thermal cut out over temperature protection device (if fitted, will be embedded in the windings).
- The protection device (temperature transmitter) on the mechanical bearings (if fitted).

The motor is either overloaded or under rated. More than likely it will be the former, especially if the motor has been running normally and within normal parameters over a long period of time.

There are many factors that can cause the motor to draw excessive current, over heat and cause electrical windings to burn out (insulation failure). Check the motor's resistance across its windings and verify that they are balanced. The common rule states that insulation life is cut in half for every 10°C of additional heat to the windings.

As an example, if a motor that would normally last 20 years in regular service is running 40°C above rated temperature, the motor would have a life of about 1 year. The following, are some examples that may lead to premature aging of a motor and ultimately failure:

- Over voltage.
- Mechanical overload.
- Mechanical seal failure.
- Mechanical bearing failure.
- Inadequate cooling on the motor frame itself (poor cooling due to high ambient temperature, clogged ducts, etc., are typical examples of nonelectrically induced temperature stress on both the motor and insulation system).

It can also be on the fan side:

- Mechanical overload.
- Mechanical seal failure.
- Mechanical bearing failure.
- Foreign bodies on the fan.
- Belts (worn or broken) and pulleys (improperly aligned) which are driven by the motor.
- The fan blade itself may be imbalanced or loose on the shaft.
- A fan may have the same physical size, but if a different heavier type metal blade is fitted, it will have a serious effect on the motor load itself, driving the electrical current load upwards.

If the fault finder can isolate/disengage the motor from the fan, it will help exonerate the fan as being the problem if the motor is still running hot.

The main questions to ask are:
Is the motor circuit **tripping**? (i.e. breaker, thermal overload, motor protection relay or VSD tripped). Is the control circuit to the control equipment of the motor itself **interlocked** with a **safety device?**
E.g. A differential pressure switch is fitted across the fan which measures the air pressure differential between the suction and delivery side of the fan itself to prove the fan is spinning. If there is no pressure difference detected on the pressure switch after starting up the fan motor, normally after a period of 10 seconds, an alarm will be activated to shut down the fans control circuit which will in turn stop the fan motor. All these **trip devices** must have at minimum, alarm indicator lights with audible alarms attached to them to ensure the alarm is properly identified, recorded and should remain active until accepted and reset by the maintenance personnel.

In other words, is it a **motor control circuit issue** or a **safety device issue?**

If it's a motor issue:

1. The supply voltage (400V) and the full load current (FLC) of the fan motor must first be identified from the motor nameplate – e.g. 400V/45 Amps.

2. The voltage must be measured with a Volt meter, a reading of 400V should normally be expected, respectively, across all 3 phases. The current must be measured on all 3 phases going to the motor, with a clip on ammeter. Normally, if all is OK with the windings of the motor and associated connections, the current readings should be evenly balanced across all 3 phases. The actual running current of the motor will be dictated by the mechanical load it is driving. A higher current usage could be indicative of a defective motor bearing and is **why it is vital to record and document the base current load of a new motor installation for reference in the future.** A motor driving a full load could possibly use up to 100% of its full load current on its nameplate when fully operational.

3. Heat and poor ventilation can have a very detrimental effect on a motor and may not manifest itself for hours after the motor is operational. Has the motor got PTC thermistor protection? i.e. as the temperature in the motor windings rise and reaches the PTC sensor's temperature rating (which is embedded in the motor windings), the PTC sensor's resistance transitions from a low to a high value which triggers an alarm. The set point level which is set on a thermistor monitoring controller whose relay contact is wired in series with the motor's starter control circuit, if activated, will shut down the motor itself. The thermistor controller galvanically separates the thermistors in the windings from the motor's starter control circuit (i.e. no physical contact between the 2 circuits).

4. Remove any accumulated dirt from the motor frame by wiping, vacuuming or brushing and ensure the motor's air passages are clear. If a lot of dirt has accumulated on the motor frame and is not cleaned, this will lead to the motor running hot because the air passages are clogged up reducing cooling air flow over the motor frame. The heat will reduce the insulation life of the windings and will cause motor failure. **If the dust is conductive, it may be the cause of an explosion.**

5. If it's a Variable Speed Driven (VSD) motor then the electrical drive that monitors the motor will show the actual fault if any has occurred. VSD's normally have a thermistor protection circuit built into them where there is no need of a separate thermistor controller i.e. the thermistor in the windings are wired directly into the VSD itself.

6. **No alarm condition should be automatically reset.** It must be a manual operation and the fault/trip must first be identified, recorded and rectified to prevent reoccurrence before the motor can be brought back on line.

7. The manufacturers' motor and fan construction arrangement drawing and the 'As built' electrical and mechanical drawings must be available. This information could help identify what could possibly be shutting down, interlocking or tripping the fan motor.

Tip: Bearings must be lubricated/replaced when scheduled on a PM or if they are noisy or running hot. Excessive grease and oil creates dirt and can damage bearings. **Do not over-lubricate.**

N.B: A motor breakdown can cost **€10,000** per hour in downtime if it happens in a paper mill or another continuous process. It is vital to install a high quality energy efficient motor.

E.g. a high quality energy efficient 55kW motor costs in the region of €3,000. If it's a low quality motor, it will cost a bit less. In a 24/7 continuous process, the high quality energy efficient motor will use up its own cost in electricity during the first 30 days of operation. The low quality motor will use more energy. Having been built from lesser quality materials; it is less able to convert electricity to mechanical energy and will instead produce more heat. Not only does it cost more to run, the heat will eventually start to break down the insulation materials inside the motor. This in turn may lead to a motor failure, with the ensuing downtime costs as well as the costs for a new or rewound motor. Meanwhile, the high quality energy efficient motor keeps on running.

Think before making a purchase, do not compromise on motor quality, top of the range process performance motors use the best materials and manufacturing practices to minimise energy consumption and are designed to achieve a possible service life of up to 30 years.

Written corrective procedures such as below **are a must.** A trouble shooting guide should look something like this, with individual possible faults and actions listed per instrument, valves etc. insist on it from the vendor:

E.g. **Fault: Electro pneumatic actuated on/off Nitrogen valve did not open when commanded.**

Procedure in case of fault: Immediate check and/or repair.

Possible cause	Corrective measures
Electrical or cables defective	*Check electrical connections and cables* *Check is correct supply voltage present* *Check solenoid valve coil*
Control air supply	*Check is compressed air present* *Check compressed air connections* *Check compressed air pneumatic actuator*
Proximity switches defective	*Check switches; replace if necessary*
Nitrogen valve blocked	*Check movement of Nitrogen valve* *Clean Nitrogen valve if necessary* *Check Nitrogen valve bearings and joints*

Try to create a troubleshooting checklist for each piece of equipment on site. Should a problem occur, find the problem on the checklist and take the indicated corrective action. If such a checklist does not exist, it must be created. Use the vendor who supplied the equipment, peers or colleague's knowledge of the equipment, collate and populate the information into one document that everyone can use.

No trouble shooting document is ever written in stone, consider it to be evolving. As a person comes across unforeseen issues and rectifies them, ensure this new information is captured where it can be referenced so others don't have to cover the same ground as the original fault finder. This saves valuable time and money to the company which could be better spent on more constructive issues.

A person must be prepared to accept that there will be 'Highs and Lows' during their career.

The 'High' of overcoming and rectifying a serious fault, putting the plant back online, saving the company vast sums of money is very gratifying and also gives a sense of deep confidence in one's abilities.

The 'Low' will also be experienced. There will be a time when it is believed all the bases are covered, but an unforeseen problem arises which can neither be diagnosed nor rectified. A person may feel they have lost 'face' in front of their superiors. Doubts may start building up on their abilities and fault finding methods.

Put it down to experience, learn from it and move on. Keep this in mind, when the 'Lows' in a profession are experienced, if dedicated to a discipline, professionalism will make sure these 'lows' are kept to a minimum.

Hindsight is a great thing, having diagnosed a fault which may have taken a person hours to pinpoint and rectify. Then having to listen to their peers or the 'Monday morning quarter back' telling them how they would have resolved the fault in minutes.

It is easy in the cold light of day, under normal circumstances, to discuss what one would or wouldn't do when a fault develops unexpectedly. Rational thinking will prevail.

Words are easy to say where a list of possible causes might be rolled out by a number of people. It is a lot harder in the real world to actually prove each suggestion. Ultimately, one of the suggestions might be right and then have to listen to 'I told you that was the issue'. This type of fault finding is too 'Hit and Miss'. Stick to proper fault finding methods no matter how tempting it is to **'Chance it and see what happens'.**

Check, double check and treble check any work carried out. Be satisfied that all work completed was done correctly without compromising safety and to the best of one's abilities, regardless, of the time pressures they may have been under to get the piece of equipment back on line.
One of the tests of being a good engineering practitioner is the ability to recognise a problem before it becomes an emergency. As the qualified person, it's their responsibility to make sure everything is running like a well-oiled machine. In doing so, they must keep an eye out for any potential threats. Have the ability to **'think on their feet and adapt to the situation'.** Determine what the best solutions are and act quickly and efficiently if problems arise.

In extreme circumstances, if under pressure, tired, exhausted and time is critical in rectifying a fault, sometimes rational thinking may not be employed. There may be only a limited amount of time available, possibly minutes, **for critical decision making to be carried out** and to ensure the decisions made, are the right ones.

E.g. You may find yourself in a highly charged atmosphere (you are the ships' engineer, out at sea, the engines have failed, the ship is listing and dead in the water and there is a violent storm coming in, 20 lives plus the ship and it's cargo are in your hands). Taking the wrong course of action, could possibly exacerbate the problem.

Running around throwing your hands in the air and shouting obscenities is not going to help. This isn't a time for covering your back, finger pointing or the blame game. Remember, apportioning blame is not constructive, look for ways to resolving the problem instead.
'A person must remember their training', remove outside influences and distractions from their mind, stay calm and focused. Investigative meetings of what went wrong can be held at a later date once the issue has been dealt with. Preventative measures can be agreed to avoid reoccurrences.

TV documentary channels on industrial explosions, fires, plane crashes etc. are very informative where one might think that this will never happen to me. **'Think Again'** (70% of reported incidents in the oil and gas industry worldwide are attributable to human error, accounting for in excess of 90% of the financial loss to the industry). A person may work in the Nuclear, Aeronautical, Chemical, Petrochemical industry where exothermic (process that generates its own heat) reactions take place.

E.g. As the shift maintenance engineer, you may be faced with a situation where the safety equipment designed to mitigate the consequences of an emergency has failed. The entire cooling system (including back up cooling system) fails during an exothermic chemical process.

A critical situation has now arisen, the process temperature starts to rise exponentially with a risk of explosion and the cooling systems need to be put back on line as soon as possible. **'Delegate'**, if there are other personnel on site, ask them for assistance, whether they are technical or not. Let them take the phone calls from other personnel warning of impending dangers. They must be told to write down all information accurately, who the person is and what phone number they can be contacted at. If the calls been received are only symptoms of the cause been worked on, stay focused on the task. Being bombarded with non essential information can be very distracting.

Deal with the 'symptom issues' systematically according to their criticality once the actual 'cause issue' has been rectified and the cooling is back 'on line'.

N.B. Never put yourself or other staff at risk by going into a building where there may be a possibility of an explosion, gas leak, flooding etc. actually occurring. Get out of the area immediately and alert the emergency response team who are professionally trained to deal with such critical issues. Remember, in the space of seconds you can be overcome by fumes or electrocuted should there be water streaming down into an electrical panel. You may be about to enter into an area where you see another member of your team lying on the ground. **'Stop and think before you enter'**. Always be aware of the dangers around you.

Personnel may enter a control room of a plant where there are numerous flashing lights and alarm sirens ringing in their ears, warning of some impending danger unknown to them. This can add complexity to an already highly charged situation. They might be receiving misinformation from the instruments that control the process which feed into the control system or from the control system readouts themselves. There may be multiple erratic alarms going on and off. Can they trust them? Personnel must try to analyse conflicting information and try to decipher which alarm is real and which alarm is false.

Don't panic, panicking will blur a person's thinking. Ensure not to misread the information being received from the current environment. Review the standard operating procedures; ensure all settings on the control panel are as they should be. Has a valve been opened or closed in error by personnel during operational checks? If one of a multitude of instruments on a control panel is malfunctioning or giving questionable readings, perform a primary scan of the entire control panel and check all instruments and their respective readings. Are any other units not working properly or giving the same type of erratic readings? Go through them one by one, starting with the most critical instrumentation readings.

Flexible behaviour will play a key role in this situation. Personnel must have the ability to adjust their thoughts to changing situations and conditions. Adapt and adjust their thinking to new information. Flexibility involves being able to train oneself to reinterpret unexpected situations that may at first give a sense of doom or alarm. Flexible people have the capacity to smoothly handle, multiple demands, shifting priorities and rapid change. Be ever mindful of 'task saturation' **e.g.** trying to do too many things at one time. Some critical action on a check list might be missed, which could exacerbate the situation. If a person can't assess what's going on in their environment, they'll have difficulty adapting their responses to this type of 'real time' information.

If there are others also trying to solve the problem, don't suffer from 'collective brain thinking' **i.e.** everyone focussing on one issue, which may be only a symptom of the actual cause. Don't get caught in this trap, all involved must take in as much information from the environment as possible. Disseminate it and make decisions based only on accumulated **accurate** information.

'Mental Gymnastics' may start to kick in, personnel know what should be happening, yet the exact opposite is occuring. One must have the ability to adapt to unfamiliar, unpredictable and fluid circumstances. Try not to become confused and make rash decisions. Remember fault finding discipline must be adhered to. ***Manage the problem.*** **Try to understand exactly what is to be done and the consequences of those same decisions before any actions are taken and ensure all decisions made and actions taken are recorded to be reviewed after the crisis has passed.** The first objective is to get the piece of equipment stabilised and in a safe operational mode i.e. figure out what is controllable, what is operational and not operational.

Pressure may be coming from superiors for answers, don't make statements just to appease others which may prove to be totally fact less. Being a hero to your superiors for trying to get the plant back online quickly, may sound great. Keep in mind though, if short cuts are taken or a safety system is bypassed without fully understanding the ramifications of such activities and the process becomes even more unstable and explodes, the 'Buck will stop with the person who carried out these uncontrolled activities'.

N.B. Only propose solutions when the 'cause & effect' of any actions carried out are fully understood.

Personnel must stay focused, calm and remember their training and above all. Give answers, based only on accurate engineering fact.

These cases may sound extreme, but, all events actually happened. Hopefully the reader will never be confronted with any similar type events, but having been trained properly and to avert disaster, the feeling alone from this will be very satisfying.

The above examples stress the importance of knowing the plant's systems inside out, proper training, work practice and preparation on how to handle critical situations. Imagine been in this serious situation and not having the proper training and not knowing what to do, lives, including your own, could be lost as well as severe damage to the plant. The costs associated with such an accident can be astronomical. The environment and the company's reputation could also be affected. **Adequate expert training must be given.**

A person cannot be expected to trouble shoot a problem on a piece of equipment, if they have not been taught or trained how to deal with a problem in the first place or if inadequate alarm systems warning of impending danger or systems operation failure are not installed. **Insist on proper training.** Your life or those of colleagues could depend on it. Asking a person to self train on a complex piece of equipment is **not recommended,** as this person cannot train properly and be competent to work with this equipment when **'they don't know what they don't know'.**

In any industry, whether it is a manually intensive production line or a highly automated production plant where product output is paramount and making money may be the dominant factor, **safety** sometimes, can take a back seat. Watch out for hazards, identify, report and rectify them. Don't wait for an accident to happen. Even if the piece of equipment is running without any operational problems, it doesn't mean it is running safely.

Regardless of the time constraints or how much pressure a person is put under by superiors to get a piece of equipment operational, **they must not compromise on their experience, knowledge and know how.** <u>If they know it is not safe</u> to put a piece of equipment back into service, **they must not** let a superior over ride their decision. Ensure any concerns are documented and followed up if decisions are overridden.

N.B. Use 'Best Engineering Practices' across all disciplines. It should become the exception rather than the rule **not** to use these practices.

Part and parcel of maintaining a plant will be written checks. These will be carried out on pieces of equipment throughout the site using GDP (Good Documentation Practices). A person must clear their mind. Make a conscious decision to take their time. Write neatly and fill in the correct information. Leave no boxes unfilled and ensure to sign and date all written information carried out.

E.g. When writing a cheque, the currency amount is filled in, crossed, signed and dated, the same goes for a standard maintenance check.

These checks can prove very useful should a problem occur on a piece of equipment. The archiving of this information can assist in fault finding if a subtle but changing pattern manifests itself on the equipment over a period of time.
N.B. These records are also vital should an accident or incident occur. This is the first document (maintenance records) that will be viewed by incident & accident investigators.

A person's signature is their bond and by signing a maintenance check and approving it, they are signing that everything is in order, the equipment is within acceptable running parameters and is suitable for operation.

N.B. Never sign your name to a check that you didn't witness or carry out.

In the technological world we live in with computer screens and there associated control systems now operating plants, it is easy to sit in a control room and assume the control system that operates the plant has everything under control. 99.9% of the time, these systems work flawlessly, but, if erratic readings are showing up on the computer screens (i.e. unexplained high or low temperatures, pressures, flows etc.) Go on plant if possible and physically check the reason for the erratic readings **Also watch out for flat lines on data trends on PC screens where there would normally be slight oscillations.** Flat lines can be indicative of a faulty probe, open or short circuit on the control cable between probe and control panel, blown control fuse or faulty PLC input card.

Don't fall into the trap of always taking for granted and assuming that the computer read outs are correct. Automation systems are programmed and can only react to known operational and safety conditions that have been envisioned by the engineers during the initial design. The automation system then acts appropriately to deal with the issue. It can do nothing about an unforeseen condition that was not anticipated by the designers. A condition such as this can have a devastating affect on a piece of equipment's safety and operational system and will not be detected by the automation system, as it is not programmed to do so. Don't let flippant remarks from others such as, 'It always does that' satisfy curiosity, **investigate and report.**

Modern control systems have fail safe systems including, duplicate watchdog systems, duty/standby, redundancy and self monitoring hard wired critical variables of the equipment which are flagged, alarmed and actioned by the control system that monitors them. Never underestimate the human factor in any control system, i.e. **to actually physically check any fault on plant regardless of how inaccessible a piece of apparatus may be.**

Always insist on the best **'turn key project'** installation possible e.g. the more instruments that are fitted on installations, the easier it will be to keep the plant running efficiently and if a fault develops these instruments can help the fault finder to diagnose the fault quicker.

Remember, a control system is only as good as the instruments that control and monitor it. Data acquisition is essential. The key is to maximise uptime and this is best achieved if you are armed with pertinent machine data.

Unfortunately, when a project installation budget is tight and money is running low, the first cutbacks will always be made to the non essential equipment i.e. 'the nice to haves'.

The failure to design a system properly is the same as designing for failure. Remember, no amount of maintenance can overcome poor design.

Always have the justification facts to hand when questioned by project engineers, finance managers etc. Insist on the installation of vital equipment and extol the benefits of them in the long run.

The time given to writing a good URS and FDS, be it a small project involving 2 weeks work or a new installation which may take 12 - 18 months to complete cannot be underestimated. Try to get it right first time. Dedicate as many meetings as it takes with the relevant personnel who have crucial experience, information or abilities. All are important criteria for a successful project outcome.

It is important to have a team of people who between them understand the entire operation of the business. Get the relevant departments respective input on the engineering, operations, environmental, health and safety, financing, quality etc. This in-depth knowledge of the business is critical to ensuring the processes established are going to be practical for everyday use. It is vital to have diversity in the group. Different view points about how the goal is attained will surface and various approaches should be encouraged. This removes the 'we should have thought of that' scenario after the project has been completed.

After the project is finished, call another 'lessons learned' meeting and document the **successes** and if any, the **shortfalls** that may have occurred i.e.

Successes:

1. URS & FDS were excellent which made the PM relatively easy.
2. Project was completed within budget.
3. Project was completed before time which left extra time for testing.
4. Bonuses were given to personnel for excellent work and early delivery of project within budget.

Short Falls:

1. **Bad Project Management** - poor design day one – wasn't designed or planned properly, not enough 'know how' or time was given to this phase of the project.

2. **Project not done on time** - unrealistic time frames.

3. **Project ran over budget** – unrealistic budget figure, not enough or no contingency allowed for, rising labour and material costs not considered.

4. **Personnel issues** – accidents, sickness, strikes, inexperienced people hired with little or no project installation experience (may have the qualifications, but very little experiential know how), key personnel leaving for other jobs

5. **Weather** – poor weather conditions, personnel unable to work.

Learn from successes, but, just as important, learn from the shortfalls. Put the shortfalls down to experience and try to ensure they do not happen again for future projects.

Fault Finding Overview

- Always be proactive in failure prevention.

- It is critical you understand a piece of equipment's 'Principle of Operation', i.e. how it works.

- There are many different types of faults, from minor to major. Ensure any action you carry out does not exacerbate the fault.

- The extent of preparation should be directly proportional to the size of the problem to be rectified.

- **Do not** make whole scale changes, take a structured approach and take one step at a time.

- Don't start changing parameters (P.I.D settings) in controllers, pushing buttons, turning knobs, closing/opening hand valves etc. if you are not positive you know what you are doing.

- Always exercise caution. Use logical and deductive reasoning at all times.

- An intermittent failure if left unaddressed is likely to 'crop up' over and over again until finally, corrective action is taken to rectify.

11

Spare Parts

A typical fully automated machine will normally have all the latest technologies attached to it. If any part fails and stops it, a full inventory of critical spare parts must be available.

Training and knowledge are a huge advantage. Being able to diagnose a fault quickly is key. But if in the event of failure, the spare parts are not available to perform a quick turn around, the machine will remain inoperable until the vital part is delivered.

Try to standardise on pieces of equipment. This will keep spare parts to a minimum and avoid keeping costly pieces of equipment sitting on shelves in the engineering store.

Ensure a proper, well maintained inventory of critical spare parts exists on site for all equipment that is vital to keep the plant fully operational at all times. A set quantity of all critical spare parts (with dedicated individual part numbers) must be kept on engineering stores shelves. When the minimum amount of any particular spare part has been reached an automatic reordering of the spare part should be initiated and restocked if possible via **a dedicated computer based engineering stores management software system.**

An engineering stores management software system allows the protection of a company's investment in plant equipment. This type of system will ensure a company manages its engineering inventory. Downtime will be reduced, engineering costs will be kept under control and investment in engineering spare parts will be minimised by having such a system.

Key Benefits

- Gain a clear visibility on spare parts and consumables usage via electronic stock management system bar-coding and automatic Email communication process guarantees stock levels are upheld at all times.
- Facilitates the processing of all transactions accurately and efficiently i.e. less paperwork.
- All purchasing and associated costs of spare parts ordered and managed via PO i.e. no PO/no spare part.
- Analyse engineering stores effectiveness through KPI Reporting.
- Improved system information quality with key information available to other key users i.e. finance department/ external auditors.

If asked to maintain a piece of equipment or a vendor supplies a new piece of equipment, ask about the spare parts. Do they keep a minimum stock on their shelves? How quickly can they arrive on site if needed? This could save the company having to keep expensive parts on plant 'just in case'. Ask about the after sales service they provide.

Generally a company tends to get what it pays for, so when choosing equipment which will last for 5, 10 maybe 20 years, quite often it is not the cheapest supplier that should be considered but the supplier that provides the best overall value when machine specifications and support are considered.

Replacing components with non authorised parts could lead to a sharp reduction in performance, a risk of premature failure and could possibly compromise safety and quality. Always work, where possible, with 'best in class' suppliers whose products, technical innovation and support can be fully utilised when needed.

Vendors will expedite 'spare parts orders' much quicker if a customer specifies the order number of the machine part or plant. The following guidelines should be observed:

- Spare parts should only be ordered in accordance with the vendors' recommended spare parts list that accompanies the equipment.

- Orders placed by telephone to the vendor must always be confirmed in writing with a valid purchase order (PO) supplied by the customer before delivery.

- Ensure to include cost of **delivering** the actual spare part to site in the purchase order, as the associated costs can be substantial.

- If a vendor receives orders without the above specified information, they will not normally guarantee correct delivery. They will not bear any expenses which are incurred due to the need to 'exchange parts' which can be very expensive especially if it involves large items coming from overseas.

- Delivery of parts ordered in accordance with the spare parts list shall be made on the basis of the vendors 'General Terms of Sales and Warranty'.

By observing these points, it makes it easier for the vendor to process orders effectively and to provide fast delivery.

12

Facilities Management

Facilities Management represents a continuous process of service provision to support a company's core business. Developing strategies to reduce operating costs while at the same time optimising the performance of assets will be critical to ensuring business continuity. A company must try to create a new mindset among employees to take a new look at processes based on real time information and business knowledge. Implementing Continuous Improvement (CI) projects, involving the onsite personnel, who know the associated systems and processes, will play a key role in its future success.

The total life cycle of any engineering system is the most important factor to consider. This is the overall cost of the system over its entire lifetime and includes not only the initial capital investment but also the cost of installation, maintenance, servicing and energy consumption. Only by taking all of these factors into consideration and adopting a holistic approach to a new system installation is it possible to benefit from all of the potential savings.

Companies are under unprecedented pressure from rising energy costs as well as legislation and climate change. Key performance indicators (KPI's) must be to keep overhead costs under control and improve reliability, predictability and profitability of its facilities.

Statistics show that buildings (office blocks, apartment blocks, skyscrapers) consume 40% of the world's energy. With new methods of generating energy evolving and improving all the time, ZNE buildings are made possible. A **zeronet energy building** (ZNE) is a popular term to describe a building's use with

zero net energy consumption and zero carbon emissions annually. ZeroNet Energy buildings can be used autonomously from the energy grid supply – energy can be harvested on-site usually in combination with energy producing technologies like Solar and Wind while reducing the overall use of energy with extremely efficient HVAC and Lighting technologies.
The biggest challenge is while using new technologies and designs, how to make buildings safe, energy efficient and as environmentally friendly as possible without compromising human comfort.

Utility Services:
Personnel must know where all the utility services for office buildings/process plants' etc., are supplied from i.e. main electrical power distribution boards, mains water, steam, compressed air supplies, smoke/fire detection systems, fire sprinkler water systems etc. Know exactly how to isolate any system at any time, do not wait until something goes wrong or to carry out maintenance on a piece of equipment before they decide to familiarise themselves with these essential services. Remember, if an individual 'fire water sprinkler head' has been activated in error (e.g. accidentally hit by a ladder), water at high pressure will be spraying out. The water pressure in these types of systems can be > 10 Bar, at these pressures, very large volumes of water will be delivered into an area in minutes. Ensure personnel know how to stop the water flow as quickly as possible. **This could possibly involve draining the entire sprinkler system pipe work.** The water damage alone to buildings floor/wall/furniture areas and associated electrical equipment can be very substantial as well as the associated costs involved in the clean up.

Computer-Aided Facility Management:
Use technology where possible. **Computer-Aided Facility Management (CAFM)** is the support of facility management by information technology using proactive asset management tools i.e. SCADA (Supervisory Control and Data Acquisition) systems keep a plant and its associated buildings running optimally.

These types of systems also assist in controlling, day to day monitoring, data trending (historical or real time), back up and restore capabilities, early detection of potential problems and alarm annunciation for all utility services on site, which ultimately protects a company's assets. Monitoring applications help avoid incidents and minimise risk. Cost can be prohibitive installing this type of technology, but, the benefits will far out way the initial installation cost in plant uptime, reliability, efficiency and operating costs.

Facilities Department Environmental Goals

- Prevent or reduce pollution to air, land and water through programs that reduce environmental impacts and conserve natural resources using Continuous Environmental Management Systems (CEMS) e.g. ground water sampling points, air emissions points, liquid effluent discharge points.

- Review environmental objectives and targets and set guidelines for reduction of environmental impacts.

- Continuously reduce environmental impacts and conserve natural resources.

Waste Management System

Waste management is the control of materials and equipment that have become redundant and therefore need to be disposed of in an environmentally, safe and cost efficient manner. Waste management of consumables and maintenance components is a must. This can include anything from waste oil to a complete demolition and disposal of an entire machine and its associated components.

The process includes sorting, recycling, clearance, collection, transportation and disposal of waste materials according to waste disposal regulations. Waste management includes substances that are in a solid, liquid, or gaseous state, and their management techniques differ for each state.

Waste management is complex but very important and if not properly managed, can be very expensive due to the multiple varieties of waste produced by industry. Different types of waste require special management techniques. There are licensed waste management vendors who will provide a total waste management (TWM) package and will deal with handling, transportation and documentation of both hazardous and non hazardous waste.

Key Objectives:

- Work towards ZERO landfill target.
- Reduced volumes will reduce land fill & associated transport costs.
- Encourage and promote Recycling by employee's onsite.
- The greater the understanding of where the waste is ultimately going, the greater the understanding of how and why certain wastes must be packaged for transport and for easy treatment at a waste disposal facility.

Typical Industrial Facilties Systems:

- Plant Potable Water distribution system
- Natural Gas distribution systems
- Plant Boilers and Steam generation system
- Hot Water systems
- Chilled Ammonia systems
- Waste Water Treatment Plant systems
- Sludge treatment plant systems
- Purified Water systems
- Chilled Water systems
- Compressed Air Systems
- Nitrogen inert gas distribution systems
- HVAC/BMS Air-conditioning, heating and ventilation systems

- Smoke Venting Systems
- Waste management
- Foul and storm water drainage
- Bund Management
- Energy Management
- Process drainage systems
- Fire Water Retention Tank systems
- Fire Water Diesel Generator Sprinkler Systems
- Fire Detection Alarm systems
- Fire Water/Foam Diesel Generator Cannon Systems
- Plant small power LV distribution systems
- Plant Electrical distribution systems up to MV.
- Plant Diesel engine powered Electricity Generators
- Wind powered Electricity Generators
- Solar powered Photo Voltaic Panel (PVP) Electricity Generators

The fundamental questions you must ask yourself:

- If any one of these critical systems actually fails, can the 'failed system' recover quickly (with very little disruption to processes) and get the plant back on line to ensure business continuity?

- Are all the necessary procedures, duty/standby systems, onsite personnel and 'on call' contractors/vendors in place to ensure the 'day to day' routine monitoring and servicing works are carried out in a safe and environmentally compliant manner?

- Have the contractors/vendors, coming onsite, been properly vetted and approved to ensure they are suitably qualified and insured before carrying out any work?

Energy Supply Contracts:
A large **'Energy Using'** company should consider employing an outside consultancy firm to negotiate and manage its 'Energy supply contracts'.

Negotiating an energy supply contract is a complex business. Securing the best deal possible in a market with traded futures is about much more than driving a hard bargain. It is essential to understand the dynamics operating within that market to identify when market conditions are favourable to securing a contract that brings competitive advantage.

Outsourcing energy purchasing allows a company to choose exactly how involved they wish to be in the process.

At one end of the spectrum **the consultancy firm** could fully manage the entire process. Alternatively, there is the flexibility to be more directly involved in negotiations and timing decisions with the option to choose online delivery if preferred.

They will tailor the level of support a company needs to assist in-house management to secure the best value for its business. The consultants' knowledge of the electricity and gas markets and the contract flexibility available from the various suppliers in these markets is normally excellent.

The consultancy firm will use its market knowledge, activity and influence to continually track what is happening in the market and to predict what is likely to happen in the future by bringing together a full complement of industry experts from across all disciplines of energy procurement, utilisation, research and analysis. They are ideally positioned to provide more than information alone, identifying future trends and adding value to market intelligence. Any specific information or reporting requirements that a company may have, particularly with reference to gas and electricity markets, or oil markets can be facilitated.

In today's complex and volatile energy markets, companies are adopting a more sophisticated approach to their purchasing strategy in an effort to ensure that cost competitiveness is maintained in a controlled and measured manner.

13

Vendors

If a vendor is supplying an equipment package, be it large or small (e.g. a water purification package), they normally cooperate with a company's commissioning and qualification inspection team. The vendor is responsible for the provision of documentation detailing the design, fabrication, testing and pre-commissioning of the equipment before the equipment arrives on site. The vendor is also responsible for preparation and execution of SAT (Site Acceptance Test) and commissioning protocols in conjunction with the company. They must also provide full as-built P&ID's, commissioning spares, O&M and Preventative Maintenance manuals and to provide a detailed list and quotation for 2 years critical operational spares.

The fundamental question a company has to ask itself is will the piece of equipment being supplied integrate with existing onsite IT infrastructure (SCADA) or is it a stand-alone system without any consolidation of data? If the answer is no, then this means data cannot be checked or added in real-time to any computerised management system. As a result, the equipment is a little bit cheaper but there is little or no production data to intelligently manage the manufacturing.

Tip: Whether the reader is a vendor or a mechanical/electrical contractor etc. delivering a service to a factory, the quality of work provided must always be top class. Just as important is the follow up associated 'service history paper work' provided. This is vital in maintaining business continuity. Ensure the paperwork looks professional, it is printed on headed paper, fully detailed with all works done, costs involved, parts used, is totally unambiguous and open to any scrutiny. If providing reports on the status of equipment at regular service intervals, ensure all writing is relevant and easily readable with proper spelling. These reports (hard or soft copies or both) are normally scanned and sent to the customer. All service history paperwork must be stored safely and indexed properly for easy retrieval by the customer.

N.B. Customers will very much appreciate these types of vendor reports as will the auditors who view them.

A company must set up service contracts with vendors and insist on a good service package being provided at all times i.e. 24 hour cover should be considered if the piece of equipment is deemed critical and downtime must be kept to a minimum. This type of 24 hour cover does not come cheap. A decision to implement such a service contract must be weighed up against the cost or length of downtime that can be tolerated before it starts to affect plant operations.

N.B. A good maintenance contract program on any piece of equipment is the key to:

- Ensuring that the safety features are in correct working order.
- Adding longevity to its parts.
- Optimising its performance.
- Reducing breakdowns.
- Ensuring a long service life.

Purpose of a vendor's Maintenance and Servicing Instructions:

The maintenance and servicing instructions are intended to convey information considered as necessary by the vendor for the implementation of maintenance and servicing jobs or for the compilation of in-company maintenance instructions to specialised personnel. The maintenance and upkeep instructions are a summary of the most important technical data which are required for maintaining and servicing the equipment supplied by the vendor. The information supplied in the maintenance and servicing instructions should be considered as reference values which may need to be adapted and possibly corrected in line with operating conditions after C&Q (Commissioning & Qualification).

The maintenance and servicing instructions make it possible for the equipment technician to implement preventative maintenance i.e. information is provided relating to periodical replacement of wearing parts in order to avoid damage to the piece of equipment. This 'flags' the production personnel to plan for strategically controlled production stoppages for this essential work to be carried out whilst at the same time reducing process downtime.

Special operating instructions provided by the vendor will contain maintenance and repair instructions. These instructions must be given priority. The specialist personnel responsible for operation and maintenance must be completely and comprehensively informed of the contents of these maintenance and servicing instructions. A copy of the maintenance and servicing instructions must be kept directly in the vicinity of the equipment and in the engineering office.

Vendors will normally offer you 3 types of service agreements:

1. Inspection.
2. Maintenance.
3. Full service.

1. Inspection: equipment inspections carried out by the equipments' vendor service technicians provide users who perform their own maintenance work with the peace of mind in knowing that their system is operating correctly. The vendor service technician checks all main components and safety related systems.

2. Maintenance: this type of service agreement will ensure that equipment provides maximum reliability, availability and long term value retention. Maintenance components are changed in accordance with checklists by the vendors' service technicians. Main components and safety related systems are checked and if necessary, adjusted or replaced after consultation with the customer.

3. Full Service: A vendors' full service contract ensures that complex systems deliver optimum performance throughout their entire service life and retain maximum value. All maintenance, servicing and inspection appointments, as well as commissioning work, are carried out by the vendors' service technicians according to a company's specific needs. Main components and safety related systems are checked and if necessary, adjusted, or replaced as required. Covered under this 'Full service' contract also would be:

- Consumables and maintenance components including environmentally responsible disposal. (N.B. Hazardous materials must not be allowed to discharge into natural watercourses or drainage systems. All hazardous material waste must be kept separate from normal waste and be disposed of in a specialist disposal facility and in accordance with any statutory provisions that may apply).
- Scheduled replacement of parts (e.g. service kits, drive belts, motor bearings etc.).
- Spare parts essential for system operation.

Benefits of a good Maintenance Contract:
In all 3 cases, comprehensive service documentation provided by the vendor after the inspection provides further reassurance that all is in order and carried out from a health and safety regulation perspective also. Labour, journey and overnight costs associated with the 3 service agreements are calculated in agreement with the customer.

Always give advice to vendors supplying control equipment that is to be installed in new machinery being delivered to plant. Inform them of the 'site standard' regarding equipment and what control gear must be installed. Vendors will be pricing against each other and will be trying to get the installation contract. Some will try to install pieces of equipment that will do the same job but the equipment itself may not be up to the site standard already installed on plant. Be careful of this and insist on the site standard at all times.

N.B. Keep this in mind, if installing a piece of equipment that is not to site standard and causes the loss of control or escalation of consequences in the event of component failure, the person who installed it will be held accountable.

A normal O&M manual on a piece of equipment may be up to 25mm thick, full of parameters and listings. There may be only 3 pages dedicated to troubleshooting, quoting a number on a LCD screen with text. Personnel may spend valuable time trying to decipher what is actually the problem. Insist that a vendor supplies 'easy to digest' training material i.e. power point presentations, video clips, software based learning programs, computer based training (CBT) or web based training (WBT) with their equipment. This type of training material is essential for training purposes and can also be referenced in the future. Remind the vendor of the importance of the quality of this training material and its subject matter. Use it as a negotiating leverage tool when awarding a contract.

All original signed documentation during the contract vendor's regular service intervals must be kept on site for company records and for auditing purposes. The photocopies of the documentation can go with the contract vendor. These documents will outline the machines operation and the replacing and servicing of all associated machine parts. This vendor documentation must be read thoroughly by the engineering manager and any noted observations must be followed up e.g. compressed air being supplied to the vendors equipment could be contaminated with oil or water moisture which could be damaging parts of the equipment and causing premature failure of internal parts.

Initially, at installation and set up, vendor support should be very intensive, but gradually, once your technical personnel's knowledge grows and develops, less support should be required.

Many technical issues can be resolved with telephone support. Most vendors provide advice and assistance without charge. It is given in good faith, but normally without responsibility, this maybe enough to satisfy a company's production needs.

Personnel must ensure '**business continuity**' of their plant at all times. Ensure that spare parts are still available for older critical pieces of process equipment on site. As newer models come on line, the manufacturer may no longer keep spares for the older model. Be prepared to upgrade as recommended by the manufacturer. Don't wait until the piece of equipment stops and then find the faulty part is no longer manufactured.

There are important factors that need to be considered when choosing a supplier and some of the most important factors are listed below:

- Do they manufacture the equipment or just import it.
- 'Dependability of Supply', i.e. if they import equipment, how secure and long standing is their source of supply for future spare parts.
- Are spare parts kept on the shelf in your country or do they have to come from abroad?
- Length of time in business, bearing in mind some vendors may have previously been bankrupt.
- How many service engineers are directly employed by the vendor, remember, any warranty is only as good as the vendor.

- Do they provide accurate and timely technical information when required by the customer?
- Are they a member of an accredited agency?
- Ask the vendor to provide references from their existing 'happy' customers i.e. has the vendor a good track record in addressing customer-specific requirements.
- What is their level of 'repeat order' from existing customers?

Tip: Try to set up engineering contacts with other companies who have the same equipment/machinery, so that information can be shared regarding a machine's operation, serviceability and reoccurring engineering issues.

Remember, a guarantee is only as good as the vendor who supplies it and this can vary widely from vendor to vendor. Also, ensure any warranty includes parts i.e. 'Return to factory' - Vendor site visits are chargeable. Make sure the vendor has service engineers on the road, which means that should there ever be a need to call upon a warranty, a company can be assured of a **fast** response, typically same or next day. This is a key requirement when considering a machine supplier because a company will have bought a machine to do a specific job and so if it does break down for what ever reason, they do not want their production processes stopped for an extended period of time while waiting for either a spare part or an engineer.

Backup and support is important because a company needs to have absolute security that if an issue arises of any description that it will be resolved quickly. Downtime costs money and possibly lost customers so a company needs to have the peace of mind of on site backup if required as well as access to locally held spare parts. Think about the company's future needs, don't limit focus to today's business or today's problems. Purchasing a piece of equipment is often a 5 to 10 year decision, so the vendor the equipment was bought from is often as important as the equipment itself.

All service provider companies must supply the level to which their employee's are trained, to prove competency, before they come on site, to complete any designated specific task. The vendor will normally provide a good level of detail and put a lot of effort into compiling a training record for each of their employee's and should have no problem supplying this type of information. Any reputable vendor will see training and the upkeeping of the respective training records as being very important to maintain compliance and win new business. This vendor information must be, held and updated, as required by a company's HR department and if needed, be presented to an auditor to prove that the vendor is fully trained and competent to carry out the specific work e.g. Purified Water or HVAC systems in the pharmaceutical industry.

Tip: There are vendors who will provide what is called 'Predictive Maintenance', they will go online, download data (possibly from the vendors headquarters in another country) from on-site equipment via a telephone modem on a regular basis and analyse it. The vendor will then give their expert advice and recommend when a service is needed or when parts are starting to wear out, thus avoiding unnecessary downtime.

Main Points:

- Ensure vendors come to site with the equipment needed to complete their specific tasks. The vendor must make sure all relevant spare parts are confirmed and accounted for on site before their arrival or they bring the spare parts with them.

- Ensure to receive a vendor's report or any other relevant documentation of confirmation, on works done or to be done before they leave site.

- Make sure to understand the vendors' written report (if anything is not clear or is illegible, ask the vendor to explain or rewrite). Ensure any recommended follow up actions are carried out as soon as possible.

- This report must be filed in the piece of equipment's maintenance history folder for future reference.

14

Breakdown/ Escalation Procedure 'Out of Office Hours'

Always have a good escalation procedure in place should a problem arise if onsite personnel cannot resolve. If they are '*the last line of defence*' and have no one to turn to on plant, **they must be prepared for any eventuality,** make sure all essential relevant vendors names, phone numbers and critical site information documents are updated regularly, are easily accessible (kept in at least 2 known safe locations i.e. main security office, main archive room) and available at all times to relevant personnel. This will limit downtime dramatically as opposed to running around trying to find them during a crisis. Having this phone list accessible to all will help others resolve problems quickly as well.

A shift work pattern (2, 3, 4 cycle shift) may exist on site. When shift personnel cannot resolve a problem themselves, an escalation procedure should kick in, which may involve making calls 'out of office hours' to managers, vendors etc. When shift personnel decide to make these calls, they normally are looking for 'Technical Advice' or clarification on an issue that needs to be addressed immediately.

It is vital that the people they will be calling 'Out of office hours' are technically qualified and trained to give the shift personnel better direction, clarification or guidance in a major breakdown or incident. Avoid 'comfort calls' i.e. covering one's back phone calls.

The shift personnel on site, who cover the plant 24/7, must ask themselves the questions below before they approach the shift supervisor:

1. Is the call warranted?

2. Why couldn't they resolve the breakdown issue themselves – Specific questions: Have they been properly trained? Do they need more training?

3. If a contractor is providing cover for a person who is on sick leave, on holidays etc. Is that person qualified and can that person work safely in the environment he/she is put into?

4. Is the proper information (FDS, P&ID's, electrical, mechanical drawings, O&M manuals) readily available, easily accessible and has it been used by the fault finder in trying to locate the fault before approaching the shift supervisor?

5. Are the spare parts available if needed and ready to install if a piece of equipment had to be replaced?

6. Have the proper channels of communications been adhered to during the escalation process. Has the shift supervisor asked the appropriate questions before contacting a person off site?

The shift supervisor (as well as the engineer/technician/operator) who escalated the issue and made the phone call to the 'Manager on Call/Vendor) must be held accountable for their decisions/non decisions and actions or non actions. They must submit a written report to their respective manager, outlining the reason why they had to make the phone call, referencing that all above points had been questioned and clarified by the covering shift personnel before the call was made.

N.B. The shift supervisor should go, when possible, to the piece of equipment with the shift personnel where the fault has arisen and ask to see the problem first hand.

Example of a shift supervisor's questions to personnel before calling a person 'out of office hours':

1. Have the shift personnel been adequately trained and signed competent on the piece of equipment they are working on? If not, it must be written in the report, for the relevant management review.

2. Have they referenced the relevant FDS, PFD, P&ID's, electrical, mechanical drawings, O&M manuals for the faulty equipment and if not, why not?:

Maybe:

a. They can't find them (not readily available or easily accessible),
b. They don't know how to read or decipher them.
c. They just haven't bothered referencing them in helping to diagnose the fault.

3. Ask them their findings and how they came to their conclusions thus far. Ask them are they basing their findings on 'gut feeling' or Good Engineering fact? i.e. referencing point 2. Write the conclusions down on the report for management review.

4. A copy of this report/phone call must also be sent to the HR department to keep them aware of the issues so they can be addressed properly with the department manager.

Criteria that must be met for onsite personnel 'On Cover/ Out of Office Hours'

Personnel:

- Must be competent and confident in their own abilities as they are accountable for their actions.

- Must keep their discipline, no matter how much they would like to press a button, open a valve etc. *They must keep their hands to themselves.* Ensure any action taken is based on competent process/engineering fact.

- Must try not to suffer from **'Analysis Paralysis'** either (**thinking gets in the way of doing**).

- Cannot be expected to trouble shoot or deal with a problem on a piece of equipment if they have not been taught or trained how to deal with a problem in the first place or if inadequate alarm systems warning of impending shutdown or danger are not installed.

- Must try to understand and establish exactly what they are doing and the consequences of their decisions before any actions are taken.

- Must always exercise caution. Use logical and deductive reasoning at all times.

- Will have to 'think on their feet', adapt to a situation and deal with problems as they arise. Stay focused, calm and remember their training. Give precise answers to personnel, both on or off site, based only on accurate process/ engineering fact.

- Must use their senses i.e.

- Must look out for anything unusual e.g. oil stains, leaks, burn marks etc.
- Must listen out for anything that doesn't sound right.
- Must use their nose to smell anything out of the ordinary e.g. burning oil, rubber etc.
- Must get to know the noises the piece of equipment generates, check for any excessive vibrations by actual touch, this could be first sign of wear and tear of a bearing etc.

- Must watch out for hazards, identify, report and rectify them.

Criteria that must be met by 'Off site on call personnel'

N.B. Personnel on site who are making the call to the designated 'On call Manager', must have confidence in the person's abilities to lead them through a crisis.

The 'On call Manager' must be technically qualified and competent enough to be able to make effective decisions about, what to change and how to change it, during a situation. Off site personnel must be able to convey their knowledge to onsite personnel when a call is made.

The 'On call Manager':

- Should be able to organise and plan with regard to personnel and equipment in order to manage a given situation. They must show leadership, confidence and control in making the best use of their available resources.

- Must be accountable, give proper clear direction to on-site personnel and work through each issue until it is resolved.

- Must remember, when time is critical in rectifying a fault, the circumstances are extreme, personnel are exhausted and under pressure, rational thinking sometimes goes out the window, A possible wrong course of action taken, as advised by them, could possibly exacerbate the problem, this must be prevented at all times.

The key to all of the above is that accountability must be driven downwards and personnel through all levels must be held responsible for their actions or non actions and reminded regularly of same.

15

Energy Management

Energy efficiency is an ecological and economic obligation, it conserves valuable resources, reduces carbon emissions and the environmental impact – as well as the cost-related implications of carbon taxes. It also minimises operating costs for each and every business. More and more companies are trying to develop a 'green' approach to reduce the environmental impact of their manufacturing processes in reducing energy and waste.
What companies are looking for is tighter control on how to increase plant capacity and improve product quality. Both improvements can have a big impact on a company's running costs and its bottom line. With a little investment, many 'Energy Using Systems' have the scope to achieve higher efficiency through improved control, more regular maintenance and closer monitoring.

'If you don't measure it, you can't manage it'

If You Don't Measure it "Accurately" - You Can't Manage it "Accurately".

If you have an Annual Energy Cost of €3,000,000:

- If 10% Accuracy is accepted then Uncertainty of €300,000 per year is accepted.
- If 5% Accuracy is accepted then Uncertainty of €150,000 per year is accepted.
- If 2% Accuracy is accepted then Uncertainty of €60,000 per year is accepted.

Accuracy pays off!

Energy Management fundamentals:

The normal employee practice of almost every workplace whether it is in a commercial or industrial setting is that "if it's not coming out of my pocket, I'm not going to worry about it". It's only when the select few remind personnel of energy management or send an email that a bit of an effort will be made. Any responsible management team does not need to be reminded of the necessity for energy management and how it will help save the environment and cut down on the company's running costs. A company must change the mindset of all its employees into thinking about energy management, the way they think about it at home **e.g.** they don't turn on the immersion heater and leave it on all day just in case they need hot water.

A simple way of starting energy management whether at home or at work, is to gather received energy bills. Look at the amount of kWh's of energy expended in the billing period and apply a basic calculation.

E.g. : Electricity billing period:

> 01/11/10 – 30/11/10.
> Total of 30 days.
>
> Total expended energy: 3,000 kWh's
> Number of days: 30
> = 3,000/30 = 100 kWh's/day
> = 100 kWh's @ 17 cents per kWh expended
> = €17 per day.

The same principle exists no matter how big the kWh's expended on electricity, gas, oil etc. Just get the baseline energy figures in kWh's, add the associated vendors monthly distribution connection costs and it is simple mathematics from there. Where possible, feed the information into a PC and set up a run chart where monthly peaks and troughs will be clearly visible in energy usage and costs. A basic monitoring system has now been created.

BMS (Building Management System) - Technology is of huge assistance in proper energy management. It can read, trend and alarm real time energy consumption onto a PC for all to read and act on. It is costly to implement proper energy consumption monitoring systems, i.e. installing power, oil, gas and water meters. Then having to return all the real time power and flow readings from the meters to a PLC for deciphering and trending on a PC for viewing and dissemination purposes. Ensure to have the installation costs in implementing such a system versus payback complete, before looking for money from the management team. The cost of installing such a system will always be the driving factor to implementation.

If a company cannot afford this technology, **use the energy meters installed in the workplace.** Designated personnel must make it their business to find out where they are and start reading and logging the kWh consumption daily. This will give a company real time (measure the amount of kWh's expended in 1 hour and relate to plant activities) and daily data instead of waiting until the monthly bills come in the door. This log can be reported to other personnel in the company. It shows the current daily energy consumption on plant, without the need for expensive technology.

Full energy efficiency cannot be introduced overnight. It will be the result of a determined effort over time. Small energy management activities can be done almost immediately on any site **e.g.** switching off lights, turning off heating, air conditioning unit's etc. in areas during periods of unoccupancy. Putting up energy saving posters on notice boards throughout a company will also remind all staff that energy saving is everyone's responsibility.

The real focus is on the 'out of sight' large power users. The large pumps, air handling unit fans etc. with large motor ratings **i.e.** 20, 50, 200 kW's that run 24/7.

These costs can be astronomical, a simple 'rule of thumb' is:

For every 1 kWh at a cost of 11.42 cents per kW (industrial rate) of power a motor uses running 24 hours a day, 365 days of the year will cost approximately €1,000 per year. Now multiplying that figure by 200 kW's = €200,078 to run one 200 kW motor for one year. Energy efficiency of different pieces of equipment in an industrial plant is not always easy to detect without in-depth study and analysis.

Implement Energy Management plan – the plan will only be as good as the people who are implementing it, the time dedicated to it and the systems installed to monitor it. The gathering of the data, knowing what to do with it and what changes will be needed to adapt to this new found information will determine the successful outcome of the plan.

People like to see value for the money they invest. They like to see results, facts and graphs. Don't bombard them with figures. Have graphs, bar or run charts ready to show what energy is being expended now versus what was previously being expended before the energy monitoring systems were put in place. Ensure also to have all the facts and figures available for those who may want to scrutinise them further.

E.g. the amount of energy expended versus the amount of product made:

In one year, a company expended 100,000 kW's of energy for every 1,000 units of product produced.

The following year, after successful energy management implementation, the company expended only 90,000 kW's of energy for every 1,000 units of product produced.

A reduction of 10,000 kW's of energy to produce the same amount of product as the previous year with a cost saving of over €1,142, which can be further invested in more energy savings projects.

The EPA (Environmental Protection Agency) will be very interested in this type of information and a company is also showing their commitment to energy reduction and a cleaner environment.

Tip: There are Government grants available in some countries for the installation of energy management systems in industry, why not make some enquiries?

Review and evaluate – if there are multiple stand alone production plants, laboratory, maintenance and administration buildings on a company site, an evaluation and review must be done on the amount of energy each area is expending. Individual energy meters should be fitted to the utility services being provided to the individual areas thus allowing energy costs for running these areas to be charged to their respective cost centres. A great way to get company management concerned about energy usage for their respective areas is to make them **accountable for energy expended** and charge their yearly allocated department budgets directly for it. The bills received monthly from an energy provider, should exactly tally with a company's internally recorded energy figures that are being charged to the onsite energy users. If this type of control and measurement exists on site, a company is well on their way to successful energy management implementation.

Effective O&M is one of the most cost-effective methods for ensuring reliability, safety, and energy efficiency. Inadequate maintenance is a major cause of energy waste in both the industrial and private sectors. Energy losses from steam, water and air leaks, uninsulated lines, maladjusted or inoperable control valves and other losses from poor maintenance are often considerable.

In today's industry, where fuel costs are rising rapidly, to remain competitive, a company must have very little downtime and run as energy efficiently as possible.

O&M are the activities related to the performance of scheduled and unscheduled actions aimed at preventing equipment failure or decline, with the goal of increasing efficiency, reliability, and safety.

Good maintenance practices can generate substantial energy savings and should be considered as a resource. Striving constantly for cost efficiency driven by reduced maintenance, will play a key role in any company's long term business strategy. Any piece of equipment that is running at low efficiency will incur year round penalties in the form of increased power consumption and higher operating costs.

If energy management strategies are employed on all of the 5 critical utility services (Water, Compressed Air, Natural Gas/ Oil, Electricity, Steam) to any site, cost competitiveness can be realised resulting in more profits for a company. Everyone on site can play their part by being more energy conscious and switching off pieces of equipment that are not in use. Moreover, improvements to facility maintenance programs can often be accomplished immediately and at a relatively low cost.

Down time = Money: The sooner a faulty piece of apparatus can be brought back on line safely and fully operational the better. Maintenance, downtime and spare parts all cost money and can add substantial costs to the daily operating budget of the plant. Ultimately the profits of the company will suffer. Personnel must run a plant as cost effectively and efficiently as possible but not to the detriment of safety. Taking risks or shortcuts may lead to serious injury or damage to plant.

Regular scheduled maintenance ensures equipment is kept in optimal condition at the lowest overall operating cost. Improvements in energy efficiency can be made in electronic design by using energy efficient components, such as microcontrollers or up-to-date power management systems.

Energy costs can be addressed through the use of variable speed drive technology or intelligent motor control systems. It can also be addressed through advanced power management programmes that predict, assess and audit usage and then create plans and procedures to help optimise energy usage.

Tip: Electric motors consume an estimated two-thirds of the electrical energy used in industry. VSD's may be configured to both power a motor and during braking, re-circulate stored energy back into the mains supply, reducing energy consumption and significantly reducing energy bills. High energy costs coupled with reducing costs of VSD's are leading to faster 'Return on Investment' (ROI).

By being able to vary the speed and torque of an electric motor, and in turn the driven load, the following benefits will be realised:

Substantial energy savings – especially when considering either new installations or equipment packages, replacing oversized and underloaded motors. Rather than having an electric motor running continuously at full speed, an electric drive allows the user to slow it down or speed it up depending on the demand.

Reduced need for maintenance – Being able to vary speed and torque of an electric motor means there is less wear and tear on the motor and the driven machine. For example, the ability to bring a process up to speed slowly prevents the sudden shock loading that can damage a motor and the driven machine over time.

Did you know?

About 65% of the electricity in industry is consumed by electric motors and yet, less than 10% of those motors are fitted with a variable speed drive. If more motors were fitted with a variable speed drive, the energy savings would be enormous and also would be very kind to the environment in reduced CO2 emissions into the atmosphere due to the reduced energy consumption.

One of the biggest benefits of controlling the speed of an electric motor according to demand is the energy saving opportunity over other control methods that are used in combination with motors running at fixed speed. For example, in pump and fan applications, using AC drives can cut energy bills typically from between 20% to 50%. A pump or a fan running at half speed uses only 25% of the energy used to run at full speed.

Tip: Review the type and grade of lubrication used on motors and machinery. Using the wrong lubricant can add 5% to energy costs. Additionally, some high performance lubricants can reduce energy costs by 5%.

compressed air:

- It takes 7 kWh of electrical energy to produce 1 kWh of compressed air.

- Leaks account for 40% of all losses but are simple to control. Losses through a 4mm diameter hole cost approximately €540 every month, so a routine system of checking and repairing leaks should be established.

- More energy is needed to generate air at high pressure so the generating pressure should be reduced to the minimum level acceptable. Reducing operating pressure by 1 bar can reduce operating costs by up to 5%.

- The intake air for a compressor should be cold and dry and taken ideally from outside the building through an inlet that prevents rain ingress. Every 5°C fall in air intake temperature reduces operating costs by about 2%.

Tip: Lighting in commercial and domestic buildings uses a lot of electricity. A typical office building using T8 conventional fluorescent lamps and ballasts can save up to **60%** by:

- Switching to electronic ballasts with less losses, savings up to 25% can be made.

- Using dimming controllers and dimmable ballast in areas with natural daylight, can save up to 30-60%.

- Adding occupancy detectors in meeting rooms to switch off lights can save up to 40%.
- Switching the T8 tubes to T5 lamps can save 10-15% with typical return on investment in 3 months.

Tip: Consider using LED lights instead of standard incandescent bulbs or fluorescent tubes as an energy efficient and long lasting light source.

Tip: 'Out of sight' plant rooms in factories, offices and other such places should have push button timer lighting control with key switch override for long maintenance periods. For safety, a couple of low energy fluorescents or LED type lamps should be illuminated - 24/7/365. Replace halogen lamps with LED retrofit lamps. Typical life for LED lamps is 25,000 hours over the 2,000 hours of a halogen lamp.

16

Helpful tips

A Domestic Central Heating Radiator System

Problem: Radiator in one bedroom is only slightly warm, which is normally indicative of a circulatory flow problem and possibly caused by air locking, rust, black sludge or scale deposits which are all enemies of any heating system. A logical approach should now be implemented in identifying the actual problem itself.

Check other radiators, are they at the same temperature or hotter.
If all other radiators are hotter, the problem is localised to this one radiator. Check are both radiator valves open.

If both radiator valves are open, get a radiator key and bleed the radiator of any gases (these gases are normally a by product of rusting) that might be trapped via bleed screw (situated at the top of the radiator itself) Be careful, this venting gas or water can be very hot and can cause skin burns.

If water is present and the radiator is still only lukewarm, it could be a partially blocked local radiator valve which would have to be removed and cleaned or it may be that the hydraulic pressure of the pump is not capable of delivering the hot water required to heat the radiator. The hot water may be taking the least line of resistance, thus, bypassing the luke warm radiator, in this case the system will need to be balanced. Start throttling down the radiator valves to other hotter radiators in the system to force more hot water flow to the lukewarm radiator. If this does not rectify the problem, the pump itself may need to be replaced with a larger pump capable of delivering the hydraulic pressure needed to heat all radiators (this might happen on a new installation or if new radiators were added to the original system and the hydraulic pressure of the pump itself was not taken into account).

If you find a radiator that is hot on the bottom and cold at the top, bleed it using a properly fitting radiator key and check if water comes out (be very careful when doing this as this water is normally very dirty and can stain carpets etc., have an old cloth to hand to prevent dirty water being bled spilling onto the carpet). Check another radiator on the same floor and see is the problem universal to all radiators or localised to one. If you bleed another radiator and no water comes out, turn off the heating system immediately. Investigate why there is no water present. The first thing to do is check the heating system water head tank (in the attic) if it is not a pressurised system. This keeps the water topped up in the entire central heating system. If this tank is empty, either the ball cock is faulty or the water is turned off, check both regularly. Once water is re-established and back in the system, bleed the radiators again until all air is removed and water is present, switch the power back on to the heating system and check all radiators are heating.

If a radiator is found to be cold at the bottom and warm at the top, it is normally a sign that there is black sludge in the radiator. Close both flow and return hand valves to radiator, completely isolating it from the system. The water will need to be completely drained from the radiator by loosening the screwed fitting between radiator itself and the radiator hand valves. Have an adequate sized container available when draining it and don't allow any spillages (To remove the water quickly, open the bleed screw slightly). The radiator must now be taken outside the building, flushed with a water hose (common garden hose will do). Before doing so, ensure the open ends of the radiator are stuffed with old rags to prevent any residual dirty water draining out while taking it outside the building (this dirty water will destroy carpets if it spills on them). Stand the radiator up, remove the old rags and give it a good flush until all silt is removed. Let it drain completely and reinstall into the heating system, ensuring all fittings are water tight. Open hand valves to the radiator and allow to fill, when water comes out of the bleed valve, close immediately, it should now be fully heating. Check radiator over the next 24 hours for any water leaks or droplets and tighten fitting to maintain water seal,

N.B. 'Do not over tighten the fitting', you could possibly shear the threads, use a proper sealant such as PTFE (Plumbers) tape to ensure a water tight seal if need be.

Note: Above is a typical example of a domestic heating system, now, imagine being in an industrial setting where there is a heating system 100 or maybe 1,000 times larger with all the vessels, expansion tanks, buffer tanks, heat exchangers, associated pipe work, circulating pumps, control valves, instruments etc. being controlled and operated by a dedicated control system. Regardless of the size or scale of such an installation the fundamental fuel burning, heat exchanging and flow and return pumping principles are basically the same, just a bigger version.

Tip: If a central heating system is only heating certain areas of a house, check to see are there any motorised valves fitted which switch heating on throughout the house at various times of the year via a controller/time clock i.e. immersion/cylinder tank only (Summer months), radiators downstairs only etc. Check to see is there any indicator light on the motorised valve itself (this will prove that there is power being provided). If the indicator light is on and it is still not working, its internal motor could be burned out or the valve itself could be mechanically seized. Repair or replace accordingly.

N.B. Heat flows naturally from a warmer to a cooler place and is lost from a building through ceilings, walls and floors, it is vital to insulate properly. Heating accounts for up to 70% of energy bills in buildings and without adequate insulation, much of the energy that is being paid for is wasted.

TIP: Hot water cylinders should be fitted with a thermostat to ensure that the water is not heated more than necessary. The hot water temperature should be checked to ensure that it is maintained above 55°C to avoid Legionella and below 65°C to minimise energy use. When replacing an old hot water cylinder, install a cylinder with factory applied insulation, which will significantly increase heat retention of the cylinder.

Tip: Installing a modern high efficiency 'A Rated' boiler in a building which operates at efficiencies of > 90% can give an annual saving of up to 25% on fuel costs.

Boilers should be serviced at least annually to ensure efficient and safe operation. A poorly maintained boiler can often use 10% more energy than necessary and may also be less reliable.

E.g. A boiler temperature set point is set at 65°C and is struggling to reach this temperature although it's consuming the same amount of fuel it normally would to maintain this required temperature in the previous 12 months of operation. Have the boiler checked and serviced as soon as possible by a registered contractor, the boilers' internal heat exchanger is more than likely coated with ash deposits and needs to be cleaned. These ash deposits are preventing efficient heat exchanging taking place which leads to poor heat transfer between the burning fuel and the water being circulated through the boiler and then around the heating system. This results in more fuel being expended by the boiler, incurring higher financial costs on the property owner whilst trying to deliver the same heat output.

A boiler efficiency test including the adjustment of the air / fuel ratio should form part of each service visit.

N.B. Listen out for unusual 'boiler noises', this may be indicative of sludge or lime scale build up.

TIP: There are companies who specialise in removing undesirable scale or sludge deposits by power flushing, once the scale or sludge deposits are loosened, the unwanted debris is purged from the heating system with clean water. The system's clean water will also be chemically treated (Boiler, radiators and associated pipe work) with corrosion inhibitor to prevent further problems. This type of treatment will restore circulation and efficiency to a system.

Just like a car is serviced at regular intervals, so should a building. The property should be maintained in a 'weather and winter proof' state at all times. Circumstances outside a person's control can happen e.g. severe frost, flooding, lightning strike etc. but at least try to have prevention measures in place to protect the property that are within their control.

When moving into a new property insist from the builder or architect (before handing over any money) on having a set of schematics of all the utility runs both internal and external to the property i.e. Electricity cables, gas or oil lines, water pipes, drain pipes, foul sewer pipes, air ducts, vents. Keep these schematics in a safe place and use them as references if any further renovation is to take place or simply hanging a picture, this may save you a lot of time and money in the future.

Tip: When it comes to pipe or cable laying, whether it be an industrial or domestic setting. Take as many digital photos as possible of the exact layout of the various pipes or cables before they are 'covered in' using a point of reference for each photo e.g. a street light pole, a building, stair case etc. (it will be time well spent). Put the photos in an archive folder for future reference.

N.B. When water (which is virtually incompressible) freezes, **it expands its volume by up to 9%.** As pipes are already full with water, the pipes (copper, galvanised pipes etc.) burst due to the large pressures being exerted on the walls of the pipes and respective joints due to the water expansion itself. **The water leaking problem may not occur when the water pipes are frozen, it will occur mainly when the frozen water inside the pipes starts to defrost and begins to leak where the pipes have split or pipe joints have cracked.** Properly fitted insulation on pipes is essential in preventing such occurrences. Heat tracing should be considered, also known as **electric heat tracing** or **surface heating,** is a system used to maintain or raise the temperature of pipes and vessels. Trace heating takes the form of an electrical heating element run in physical contact along the length of a pipe. The pipe must then be covered with thermal insulation to retain heat losses from the pipe. Heat generated by the element then maintains the temperature of the pipe.

Provision must be made to protect equipment which is not self draining against frost damage in environments where it may be exposed to temperatures below freezing point.

Tip: Ensure attics and basement areas are properly and adequately insulated. Properly insulate any associated water tanks and pipe work in the areas as well. Check the condition of any metal/galvanised/copper water tanks on a property for rusting, decolourisation, pitting, or pinholes. Consider replacing with a new one or with a plastic tank. Check the condition of the associated copper pipe work also, replace as necessary or consider installing an established 'suitable for purpose' plastic pipe product. The plastic pipe manufacturer will give expert advice needed for a proposed installation. Keep 'plastic pipe joints' to a minimum where possible, especially in exposed areas. Plastic pipe will normally accommodate the increase in volume that happens when water changes to ice and then return to its original shape during a thaw, ready to accommodate further freezing. Plastic pipe and its associated fittings are suitable for most domestic and commercial plumbing applications.

Tip: If a central heating boiler is installed in an outside garage or shed, ensure a 'frost thermostat set at minimum of 2°C above zero' is fitted in the general area. This 'frost stat' will override any on/off timer and **'switch on'** the boiler and circulating pump when the ambient air temperature has dropped to 2°C, thus, attempting to prevent the boiler and associated pipe work from freezing. Also, ensure to properly insulate the garage or shed itself with good quality insulation and allow for proper air ventilation for the boiler.

Tip: Where mobile homes, caravans, summer houses etc. are concerned that are locked up and not used during the winter months, **ensure** to turn off/disconnect the mains domestic water supply to the property. Drain all water tanks and pipe work where possible, open dedicated water drain points as recommended by the appliance vendors (e.g. gas water boilers, shower mixers) and if need be, loosen actual water pipe joints to remove any residual water left in the system and then retighten. Refer to the mobile home vendors' instructions regarding **'central heating systems'** as it may have an anti freeze/water mixture which would not require a drain down.

N.B. Consider applying this tip when going on a winter holiday (at a minimum, 'turn off' the mains water valve into a property and drain the water header tank by opening the cold and hot taps and then closing them once the water is drained). A person may think that this is too much trouble and a waste of water, but consider it only takes 10 minutes to carry out this tip and may save 1,000's of gallons of water being emptied into their living room. Coming home to find a house flooded with water from a burst pipe due to freezing conditions will quickly dissipate any happy holiday memories a person may have.

P.S. If there is Natural gas supplying the cooking and heating needs, turn off the main gas valve into the property as well.

Tip: If there is no running water to a house over a sustained period of time due to frozen pipes and the lavatory can't be flushed. If possible, capture any water from the drain pipes running off a roof and redirect into a bucket, it's a good way of retaining water for flushing toilets. Pour this water directly into the toilet bowl itself. The same flushing vacuum principle will apply and remove the contents of the toilet bowl.

- When was the last time the mains water valve was checked for operation into your property?
- Do you actually know where it is?
- Is it a good quality hand valve and capable of handling severe frost without cracking and leaking?
- Does it open and close and properly isolate the mains water from your property when the valve is closed?

A simple check on this valve should be done at least twice a year and remember when the valve is turned fully open, give one half turn back the opposite way (if it's a wheel valve), to prevent seizing.

Questions you must ask yourself:

- Do you know where and what water valves are situated on your property?
- Do you know where and what water tanks (if any) are situated on your property?
- Are they labelled?
- Do you know what there function is?

If the answer is **no** to these questions, locate them, and try to familiarise yourself with the valves. Check that they open and close easily and are not seized, try to ensure you do not have to go looking for them if you have a water leak and it is dripping onto your new carpet. Ensure all individual water valves are clearly labelled, if not, label them or get someone to do it for you.

If you have natural or bottled gas coming into your property, would you know?

Where the main gas shut off valve is situated if you sensed an unexplained strong smell of gas and had to turn it off in an emergency?
Do you know or have you or members of your family easy access to the **emergency phone number** for the gas company? Other emergency 'call out' phone numbers such as electrical power supply or water supply companies should be up on a wall (in the utility room or kitchen) for all to access in the event of a breakdown of any one of the essential services to your home.

Problem: All electrical sockets are not working in a Commercial/ Domestic dwelling.

If in the event of a main MCB board RCD (Residual Current Device) trip and all RCD protected circuits (e.g. sockets) are not working, normal practice in older commercial/domestic installations in some countries is to use a single RCD for all RCD protected circuits. Sockets are on the RCD, lights usually aren't. Electric showers will have there own dedicated RCD.

Question to ask yourself:

Do you know where the main power board is in your property? If you don't, find it. Get familiar with the different MCB's (fuses in some cases, especially older buildings) and not to have to go looking for it if power is lost (main circuit breaker tripped which feeds the entire building) and are left in the dark, ensure all individual MCB's are clearly labelled, if not, label them or get someone to do it for you.

Investigate thoroughly for actual fault before attempting to reset the RCD. Observation is critical at all times, be satisfied after doing a sweep of the building that everything is ok, watch out for anything unusual.

If the RCD trips for an unknown reason and will not reset, a logical approach should now be implemented:

Tip: Instead of going around and unplugging all electrical appliances plugged into the socket outlets (some may be inaccessible due to kitchen cabinets, washing machines, dish washers etc.), resetting the RCD and then 'plugging in' each electrical appliance one by one until the faulty appliance is found. Switch off all individual circuit breakers at the main board that are affected by the RCD, reset the RCD and turn each individual circuit breaker back on, one at a time (allow up to 60 seconds between switching each individual breaker, as some equipment may not start up immediately i.e. has a 'run up time'). The socket circuit that causes the RCD to trip can be found much more quickly and the rest of the socket circuits can be brought back into service as soon as possible (especially freezers etc.).

Leave the circuit breaker that protects the faulty socket circuit in the 'Off position', reset the RCD, reset all other individual circuit breakers, all other socket circuits will now be back online, investigate what appliances are plugged into the faulty circuit and unplug them.

Reset the circuit breaker, the RCD should not trip, if it does, there could possibly be a cable problem that feeds the sockets or possibly a rodent problem (mice will eat anything). Assuming the RCD has reset, before you start plugging back in each appliance and turning each one on individually, check for any sign of damage to the cable feeding the appliance or the appliance itself. When the RCD trips again remove the faulty appliance and do not plug back into the socket unless appliance is repaired or replaced. Another example why an RCD may trip could be water ingress into the socket outlet itself or electrical appliance.

Tip: If using a coiled extension lead/cord to power an electric heater, kettle, transformer etc, ensure the cord is suitable for purpose and fully unwound from the coil/drum retaining it before using. The lead itself can over heat (burn marks or staining may become evident); this in turn will damage the cord and can also cause a fire. Some temperature rise may occur on the cord while a large power load is operating because of the heavy conduction current. An uncoiled and correctly rated extension lead does not normally overheat.

N.B. Ensure the cord **is not** plugged into the mains power socket when coiling or uncoiling the cord. Observe, at all times, that the cord is not cut or damaged, if so replace/repair immediately, e.g. a 'powered up' damaged or cut power cord lying in water is extremely dangerous and can cause instant death.

Tip: When screwing a screw into a piece of wood or into a wall plug in concrete, place a little soap on the tip of the screw before inserting, it will be much easier to screw in by hand or if a power screw driver is being used, it will extend the battery charge as less power will be needed.

Tip: To prevent a piece of wood from splitting when hammering a nail into it, turn the nail upside down, place the head of the nail in the spot where it is to be driven. Gently hit the pointed end of the nail with the hammer into the wood. Turn the nail the right way round and then hammer the nail home, the wood should not split.

Tip: Remember the old adage whether cutting wood, pipe, cable etc. **'Measure twice, Cut once'** it's always better to be looking at it, than to be looking for it.

Tip: If cutting or grinding metal with a grinder over a finished floor i.e. wood, linoleum, carpet etc. ensure the immediate surrounding area of the floor itself is protected with a non flammable cover. The flying sparks and miniature pieces of metal being generated by the grinder can actually leave permanent minute burn stains on the floor (ceramic tiles included) which can prove impossible to remove.

Tip: If living in an old property, try to find or map out where hidden electric cables, gas and water pipes are located **before** deciding to screw, nail or drill into walls and floors. This can prevent personal injury or a very costly repair job. This may include breaking open a section of a living room wall to get at the punctured pipe, repair it and then afterwards having to have it patched and plastered, painted etc., a lot of unnecessary expense and hardship because of one nail.

Tip: If you do have the misfortune of driving a nail or screw into a water pipe, to do a quick repair, wrap some PTFE (plumbers) tape on a self tapping screw and screw it into the hole to seal it, this should provide a temporary solution while getting a proper fitting to carry out a permanent repair.

Tip: How to charge a car battery using another vehicle.

If the **car battery** in your car has gone flat/dead, follow the instructions below to recharge it using another car:

Park a second car that is running, close enough to your car so a 'red and black' set of jumper cables can be attached. Use heavy duty 'jumper cables' at all times, invest in a good set, it will be money well spent. Turn off all unnecessary power consumption devices in your car i.e. lights, heater fans, radio/CD systems, heated rear windows etc. so as much electricity as possible will be directed into your car battery when being charged. Proceed then to turn off the ignition of the running car ensure to **remove the keys from the ignition in both cars** and keep them in your pocket (especially your car) as the doors can automatically lock when power is restored to the battery and you can be left with a different problem i.e. possibly having to break the car door window to gain access.

Be careful that the two cars are not touching. Identify clearly the positive (+) and negative (-) terminals on both sets of batteries in the respective cars, clear around the general battery area if need be (N.B. always have a working portable battery flashlight, a screw driver and a few old rags in your car).

N.B. Attaching the jumper cables the wrong way around could damage both cars, ensure to connect the red cable to the positive terminal and the black cable to the negative terminal, **under no circumstances should the cables be reversed.** Sometimes battery connections can be badly corroded, you may have to clean or scrape them down to get a good electrical connection between the battery terminals and the portable lead clamps themselves.

Take your time, ensure it is clear in your head what you are doing, make sure you are in a well ventilated area as a charging battery gives off hydrogen gas and is explosive.

N.B. Never allow the exposed metal ends of the positive and negative jump cables to touch/short circuit when connected to a battery (always keep them fully insulated from one another) large sparks will be generated and possibly ignite the surrounding area causing severe burns to you or damage to the car itself. This 'short circuit' can cause severe damage to the battery also.

1. Attach the (+) red clamp of the jumper lead to the positive (+) battery terminal of your car. Always be cognisant of where the other positive (+) red clamp end of the jumper lead that you are attaching to the battery of the other car is situated, preferably in your other hand **i.e. not left in a puddle of water (water conducts electricity) and certainly not left placed on the engine of the working car.**

2. Attach the opposite end of the same jumper lead (+) red clamp to the positive (+) battery terminal of the working car.

3. Attach the (-) black clamp of the jumper lead to the negative (–) battery terminal of the working car.

4. Attach the opposite end of the same jumper lead (-) black clamp to an unpainted metal surface on your car to provide a solid ground connection (refer to your cars' O&M manual to source where a good ground connection can be found). **Do not attach it to the negative (-) battery terminal of your car.** Ensure that the jumper leads will not be in the way of or get entangled with any moving parts.

5. Start the working car and wait several minutes for the battery of your car to charge, ***don't be in a hurry,*** only after this vital charging period of several minutes should you try and start your car. If the car doesn't start, check the connections between the battery terminals and the portable leads again to ensure there is a good electrical connection between them, wait a few minutes and try again. (**Tip:** The pitch or sound of the engine in the working car will normally drop when it is transferring an electrical charge to your car).

6. Once your car starts, **remove the jumper cables in the opposite order from when you put them on,** again, being aware of where the jumper leads are situated at all times.

7. Let your car run for a while or better still, drive around for 30 minutes or so before turning the engine off, this should be sufficient time to fully charge the battery. **The battery needs time to charge from the alternator if you want to be able to start the car again later on.**

If you find that it is getting harder and harder to start your car and you find the charge in your battery is not holding or is insufficient to start the car. Have an 'Auto Electrician' check it, the car electrics could have an internal fault with a component e.g. car alarm, which is drawing excess power from the battery. It may also be just time to have a new 'like for like' battery fitted. **Don't wait until it completely fails before fitting a new one.**

Tip: Car tyres.

Properly inflated, well threaded tyres on a car not alone make for a safer, comfortable, fuel efficient drive but also the tyres themselves act as very efficient water pumps when driving on a wet surface. A small water wave forms in front of car tyres when driving in wet weather, if tyres cannot dispel these miniature water waves, it will try to go over them (water is virtually incompressible) and this is why the car loses traction on the road due to poorly threaded tyres.

A car can weigh 1,000's of kg's which is constantly pressing down on its tyres, the better the threaded tyre, the safer and more efficient it is. A properly threaded tyre running at 100 km's per hour will dispel up to 6 litres of water per second away from the tyre giving the car better road holding **i.e.** the less water between the tyre and the road, the better the grip.

Inspect tyres regularly; remove small stones and debris from the threads. The tyre threads are also water channels, if damaged, partially blocked or worn, they lose their effectiveness. It is very important the entire tyre threads are kept at the manufacturers recommended depth for safe driving and are kept clear.

Tip: Tyres that are running on the road at 0.5 Bar below recommended pressure increase fuel consumption by 2-3%.

Tip: If thinking about selling a car, to add a little more value, ensure to retain all the paperwork associated with the car i.e.

N.C.T/M.O.T certificates, any works done on the car including details and dates of any service carried out. Insist on the paperwork and the associated costs being supplied by the mechanic. Having a full service record and associated original car instruction manuals to hand over will make the sale of a car a lot easier and will be much appreciated by the buyer.

17

General Templates

Trouble Shooting (Short Version)

Guidelines:

When problem solving, an organised approach must be undertaken. Sometimes it is possible that two relatively obvious problems combine to provide a set of symptoms that can mislead the faultfinder. Be careful of this possibility and avoid solving the wrong problem. Try to avoid being caught in the trap of blindly following instructions without having a full understanding of the task to be completed. The steps to be taken fall into the following categories.

- The problem and it's limits must be defined
- All possible causes must be identified
- Make the necessary corrections
- Try to establish what was the 'initiating factor' that caused the fault in the first place and make the necessary corrections.

The first step in effective problem solving is to define the limits of the problem. When a problem develops, compare all information with normal conditions. Knowledge and consistent records are the basis for avoiding the unusual.

Make a list of all deviations from normal operation. Do not rush in and make wild guesses. Use the systematic approach.

Delete any items not relating to the symptom and separately list those items that might relate to the symptom. Use this list as a guide to further investigate the problem.

The second step in problem solving is to decide which items on the list are possible causes and which items are additional symptoms.

The third step is not to alter several things at once; it may never be known what caused the problem. A person may then find you have the right answer to the wrong problem. Do not adjust settings and parameters. Do not disconnect or withdraw pieces of apparatus from a unit without ensuring they can be put back to their original state.

In solving a wrong problem a new one can be created.

Identify the most likely cause and take action to correct the problem. If the symptoms are not relieved move onto the next item on the list and repeat the procedure until the cause of the problem is found. Once identified and confirmed:

Make the necessary corrections. Keep incident reports. Learn from them and use for future reference.

NOTE: IF IN DOUBT- ASK.

CALL IN EXPERT HELP IF AVAILABLE AND LEARN FROM THEM

Engineering Goal

The Engineering Department is central to an organisation and what it does, as a group, by providing services, documents and information is crucial to people outside the department who are also performing crucial functions.

1. All tasks must be accomplished in a lean, efficient manner and at a high level of quality.

2. Be proactive in failure prevention.

3. New 'value added' opportunities must be seized as they develop.

4. Uncover problems and difficulties early, before they become major crises.

5. Share knowledge and expertise and feel committed to carrying out decisions.

6. Strive for excellence at all times and ensure that problems are dealt with and objectives met

Think SAFETY & ENVIRONMENT at all Times

If in doubt 'Stop' Think Ask Seek advice

NEVER TAKE A CHANCE

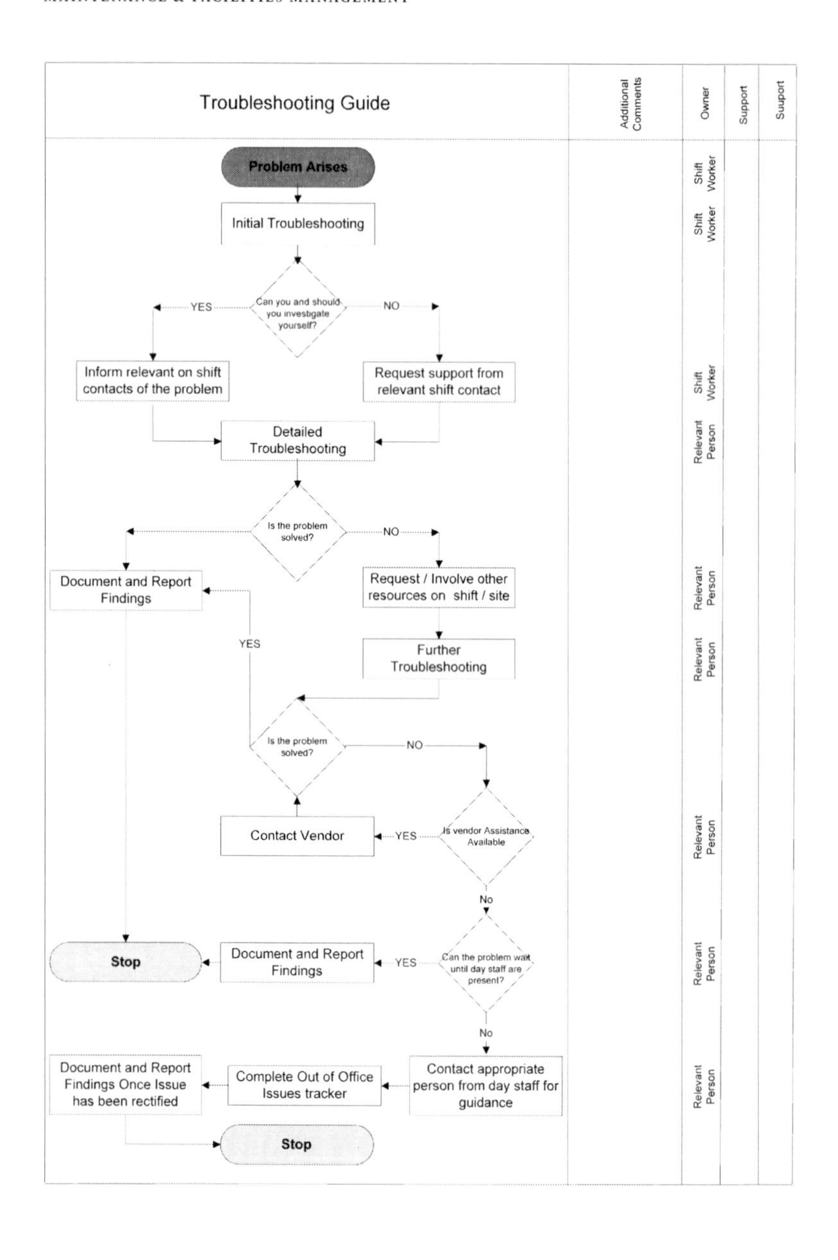

Troubleshooting Guide

Trouble Shooting Report

Name: **Department:**
Time: **Date :**

Initial Troubleshooting Questions:

1. State problem:.........
   a. What is the effect of the problem?.........................
   b. Is there more than one effect?................................

2. When did the problem start?......................................
   a. Do trends / measurements identify a time when the problem started?.......................
   b. Do trends / measurements identify a behaviour change?.............

3. Where is the problem?...
   a. Is it local to a piece of equipment?..
   b. Is it local to a step in a process / sytem?..........................
   c. Is it observed on more than one step of the process or more than one piece of equipment?.................................

Possible Causes	Facts that prove either it is the root cause or it is not

Detailed Troubleshooting

Please tick the appropriate boxes that were considered / referenced during the trouble shooting process. Please attach a copy of any relevant information found.

	Description/Comments
SOPs / WI's	
Trends via automation system	
Historical communications	
Automation documentation such as FDS	
Engineering documentation such as P&ID drawings	
Electrical drawings	
Mechanical drawings	
Technical reports	
PDA's/ Job cards	
Emails	
Other	

Findings

Write a short summary of findings

Root Cause

Was the root cause found during the shift? If yes what was it?

Follow Up

If the root cause was not identified during the shift. Write a short summary of the root cause when it is identified

Owner Signature **Date**

Review Signature **Date**

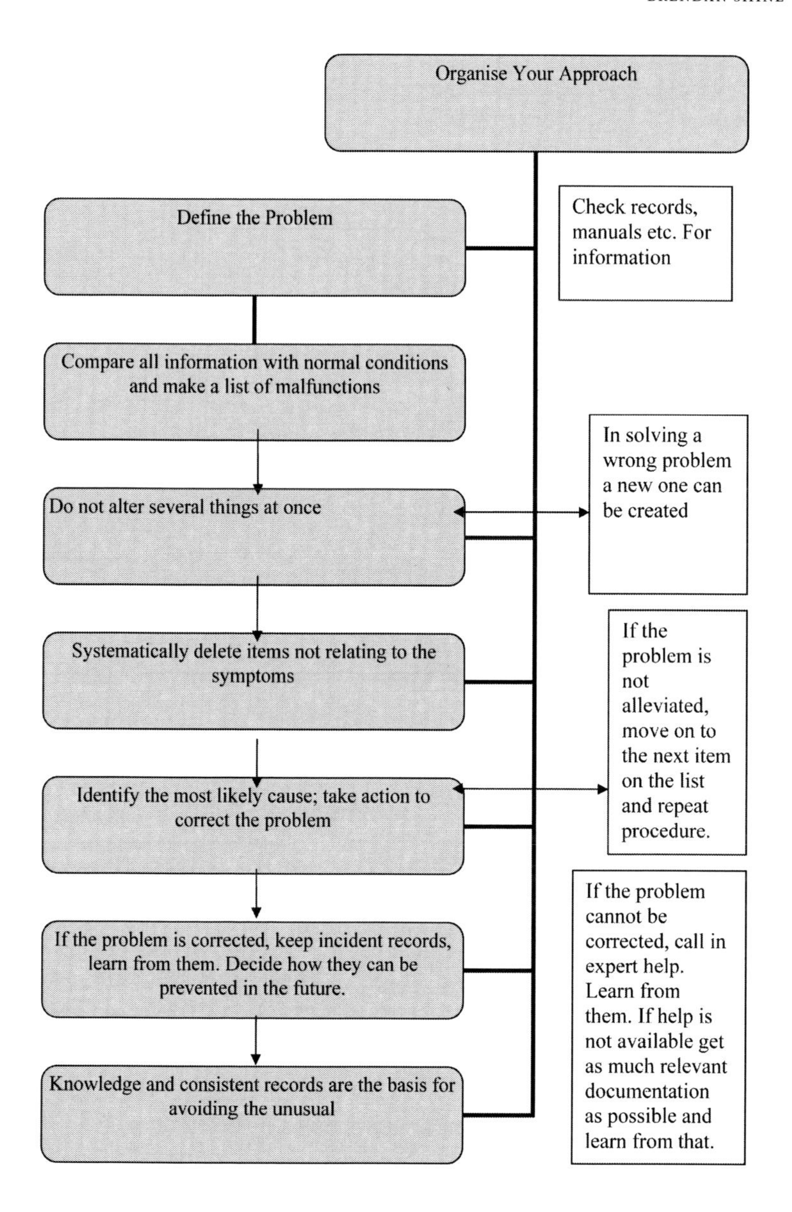

Trouble Shooting Guide (Short Version)

Curriculum Vitae - Template

Personal Details

Name	Joe Bloggs
Address	Sunny Dale, The Lake District.
D.O.B	22/11/1987
Telephone	081-1298010

Education Details

Sep 00-June 05Riverside Community College

Levels Achieved	2003 - Junior Cert passed with honours 2005 - Leaving Cert passed with honours

Work Experience

Feb 06 – Mar 10	Hillside Engineering
Position Held	Pipe Fitter
Levels Achieved	Junior Trades - Qualified With Merit Senior Trades - Qualified With Merit

Mar 10 – Present	Mountain Engineering
Position Held	Pipe fitter
Typical Duties	Reading P&I.D Drawings Fabrication and Installation of Pipe work TIG Welding

MIG Welding
Electric Arc Welding
Pressure Testing
Fabrication of Platforms
Maintenance

Possible Courses completed

- Safe Pass Course
- MEWP Licence
- Forklift Licence
- Confined Space Entry Training
- Abrasive Wheel Training
- Manual Handling Training

Relevant Skills and Qualities 'Employers' look for:

Attitude - Having a good attitude and showing enthusiasm to actually do the job is vital.

Time keeping- Being an excellent time keeper and being able to work well when it comes to getting work done 'on time'.

Hardworking- Being a good worker is a prerequisite as well as being able to 'think on your feet' and always make a conscientious effort when it comes to carrying out tasks. Being able to do things 'right first time'.

Human Relations- Having a pleasant, calm personality, and being a good communicator and interact well with other people.

Responsible- Being a very responsible person and take instruction well.

Decisive- Being able to make a decision in a set period of time. Before making any decision try to have all the facts and try to make good judgements based on the facts not anecdotal information.

Having a clean driver's license and possibly owning your own car.

Possible Hobbies and Interest

Football, keeping fit.
Computers/Internet.
Reading
D.I.Y.
Working with car engines
– building, servicing, reconditioning.

References & Referee's

18

Audit Inspection Readiness Guidelines

An Engineering audit is an evaluation of an organisation, its associated systems and processes. Audits are performed to ascertain the validity and reliability of information and also to provide an assessment of a 'plant's systems internal controls' to ensure they are compliant, transparent, effective and efficient. The goal of an audit is to express an opinion on the organisation, system etc. in question, under evaluation, based on work done on a test basis.

When a company carries out its own **internal engineering audits,** it must verify that site procedures/legislation are being adhered to and a good understanding of the hazards on plant and the range of equipment are known. A test is then performed to ensure a good understanding of requirements at all levels in the organisation exists and a confirmation that appropriate systems are in place.

A company's internal audit should do more than just check that legislation and company procedures are complied with, the following should also be carried out:

- An objective assessment of performance relative to best practice.
- Recommendations of where and how improvements may be made.
- An improved understanding of legislation and best maintenance practice by engineering personnel.
- An appraisal of equipment and associated systems with assurance that they are fit for purpose for a defined period
- A commitment to improve

An Environmental audit is an evaluation intended to quantify a company's environmental performance/compliance. These audits are intended to review the company's **legal compliance status.** Compliance audits generally begin with determining the applicable compliance requirements against which the systems and operations will be assessed.

A company's overall aim is to ensure the sufficiency and robustness of its manufacturing, engineering, and environmental systems and to have the ability to check for compliance with regulations and standards.

Always be 'Audit Ready' - **regulators can perform 'unannounced' audits at any time.** First impressions count and it's important to convey to the auditors that you have your facility (effective implemented engineering and environmental management systems) and manufacturing processes under control and **you know what you are doing.**

Remember during an audit, it is not only the company image that is been projected; it is also the professional culture as well as knowledge base that leave a lasting impact.

Having a detailed preparedness and a systematic approach for any impending regulatory inspection is a must.

Personnel must have the necessary skill set and tools to participate as a member of an audit preparation team. A company must ensure to have a thorough and effective preparation programme developed for a regulatory body audit.

Choose personnel who will be interacting with an auditor and train them on how to conduct themselves in front of the auditor.

N.B. Always ensure the general company's internal and external areas are kept in a clean and presentable state - 24/7/365 –'First impressions last'.

External 'Regulatory Body' Site Audit:

A company designate (e.g. Eng/EHS Manager) is responsible for meeting the auditor/inspector on their arrival at the plant and act as a spokesperson for the company.

The company designate is responsible for ensuring that the auditor follows the company's procedures for visitor's onsite (signing in at security, wearing identification badges etc.). The auditor must be escorted at all times during the inspection where questions are understood and addressed specifically.

Helpful hints for a company and its staff to remember during an audit are:

- Ensure the personnel who are in front of the auditors have the required technical knowledge and expertise, confidence and presentation skills.
- Personnel should always be polite and helpful.
- When a document is requested, provide the 'requested document' and no more.
- Do not volunteer information that has not been requested unless it is an advantage to do so.
- Do not guess an answer.
- Do not hide information.
- Do not argue or display anger towards the auditor.
- Never cause a deliberate delay. If for some reason you cannot deliver a copy of a document quickly, explain the reason for delay.
- Look confident and smile.

There are a couple of things that onsite personnel shouldn't do:

- Be obstructive or argumentative.
- Say something when being given the 'silent treatment' from the auditor.

- Provide answers to a question not related to an area of their responsibility/expertise, especially when their knowledge may be limited.

The auditor is not looking for faults in a system; they are looking for compliance to the standard. If a non-conformance is found, it should be viewed as an opportunity to improve, not as a reason to reprimand. To benefit a company, auditing will not only report non-conformances and corrective actions but also highlight areas of good practice and provide evidence of conformance. In this way, other departments may share information and amend their working practices as a result, also enhancing continual improvement.

Once the auditor has completed the audit, they will hold a short debriefing meeting with the company designate(s) at which they will give the company a verbal summary of the main findings of the audit. The auditor will prepare a final report of the audit as soon as possible after the audit. This report generally will set out the purpose of the audit i.e.

1. What was audited.
2. Who was present.
3. Summary of the main findings.
4. Recommendations.
5. Description of what was found and observed and recommendations.

The auditor normally sends the final report to the company. The company is then required to reply within the time frame specified in the final audit report to the recommendations setting out what it has done, or proposes to do, to satisfy those recommendations.

Assign one person to be the company's contact for receiving the audit report and answering any follow up questions that the auditors may have after leaving the site. A person should also be delegated the responsibility for coordinating any correction actions and compiling the audit report. It is important to be co-operative and to commit to providing a written audit response to the auditing body as to the audit findings.

19
Glossary

AHU: Air Handling Unit.

Air Conditioning: its principle is to absorb energy in one place and release it in another, it's designed to stabilise the air temperature and humidity within an area. E.g. a typical indoor/outdoor A/C unit set up: in the indoor unit a fan blows warm air over a heat exchanging coil where the cold refrigerant gas flows. The cold refrigerant absorbs the heat from the warm air and cool air is blown into the room. The refrigerant gas circulates through the indoor unit; it then takes the heat from the indoor to the outdoor unit via interconnecting pipework between the 2 units i.e. the refrigerant gas absorbs the energy in one unit and releases it in the other. Through compression, the refrigerant gas is heated and its boiling point increases. In the outdoor unit, the heat created through compression is released to the outdoor air by means of a fan which blows the outdoor air over a heat exchanging coil. The liquid refrigerant flows back to the indoor unit where the refrigerant is decompressed which enables it to further extract heat from the indoor unit.

A/I – Analogue Input: can have any state between 0 and 100% e.g. a 4/20 mA analogue input signal to a PLC from a flow transmitter in a process line indicates that a flow is 40% of maximum possible flow.

Analogue Instrument: if a flow rate needs to remain constant in a process stream at 40% of its maximum flow, opening the valve to 40% will not guarantee this as the upstream pressure may vary and therefore increase or decrease the flow through the valve. The way to guarantee 40% flow rate is to constantly measure the flow in the process stream using a flow meter and vary the valve opening.

Analogue loop: typically consisting of an analogue input and an analogue output e.g. a flow meter (analogue input) and a control valve (analogue output) might form a loop. The control valve is opening and closing proportionately to vary the flow rate in a process line where the flow rate is measured by the flow meter to achieve the desired amount of liquid as directed by a PLC control system.

A/O - Analogue Output: can have any state between 0 and 100% e.g. a 4/20 mA analogue output signal from a PLC is sent to a VSD which in turn varies the speed of a motor to a set speed.

Bench marking: is the process of comparing one's business processes and performance metrics to industry bests and/or best practices from other industries. Dimensions typically measured are quality, time, and cost. Improvements from learning mean doing things better, faster and cheaper.

Calibration: is the process of comparing the accuracy of an instrument reading to known standards. It is a demonstration that a particular instrument or device produces results within specified limits by comparison with those produced by a reference or traceable standard over an appropriate range of measurement.

CAPA: Corrective Action and Preventative Action

CAPEX: Capital Expenditure

CAT: Critical Assessment Team

Causality: the relation between a cause and its effect or between regularly correlated events or phenomena.

C&Q: Commissioning & Qualification

CBT: Computer Based Tool

CI: Continuous Improvement

COP: Coefficient Of Performance

Condition Based Monitoring (CbM): main goal is to increase reliability and availability of machinery, while minimising downtime, labour and repair costs. CbM can be performed by scheduling downtime, labour and materials based on machinery health. Condition monitoring is a maintenance process where the condition of equipment with regard to overheating and vibration is monitored for early signs of impending failure. Equipment can be monitored using sophisticated instrumentation such as vibration analysis equipment or the human senses. Where instrumentation is used, actual limits can be imposed to trigger maintenance activity.

Confined Space: the term confined space means any place, including any vessel, tank, container, vat, silo, hopper, pit, bund, trench, pipe, sewer, flue, well, chamber, compartment, cellar or other similar space which, by virtue of its enclosed nature creates conditions which give rise to a likelihood of accident, harm or injury.

Critical Instrument: An instrument where failure or inaccuracy may lead to an EHS or GMP non-compliance.

CRU: Condensate Recovery Unit

CTP: Commissioning Test Pack

Current - Symbol (I) - measured in Amperes - Electrical formulae: **(V / R)** or **(P / V)** or **$\sqrt{}$(P / R)**.

CV: Control Valve

DCS: Distributed Control System

D/I - Digital Input: A digital input can have only 2 states i.e. on or off.
E.g. an on/off actuated valve with feedback micro switches fitted to the on/off actuator on the valve signals to a PLC of its position – opened or closed.

Digital instrument: an on/off valve can only be opened or closed; a switch can only be on or off.

D.I.Y.: Do It Yourself

Digital loop: is a closed circular arrangement consisting of an input and an output e.g. an on/off valve with feedback (micro switches or proximity sensors fitted to the on/off actuator on the valve to tell control system of its position) forms a digital loop i.e. a valve is energised by a control system output and the valve sends a feedback signal to the control system input to confirm its open or closed status. When the valve is directed to open, the control system expects to get a feedback to say the valve has opened. This is a loop.

D/O - Digital Output: A digital output can have only 2 states i.e. on or off. E.g. an on/off solenoid actuated valve opens and closes via a digital output signal from a PLC i.e. the actuated valve opens or closes as the PLC output signal energises/deenergises the solenoid.

D/Q - Design Qualification: the purpose of DQ is to compare, using a structured approach, the proposed design with the user requirements to provide assurance the proposed system 'Design Critical' items will satisfy the needs of the system owner while conforming to all pertinent regulations.

EED: Energy Efficient Design.

E.g.: means "for example" and comes from the Latin expression 'exemplum/exempli gratia'.

E, H&S: Environmental, Health and Safety

EPA: Environmental Protection Agency

etc. – et cetera - is a Latin expression that means: "and the rest"; and others; and so forth: used at the end of a list to indicate that other items of the same class or type should be considered or included

Facts: are supported by clear, undeniable evidence and may be confirmed or disproved.

FAT: Factory Acceptance Test.

Fault tolerance: is a system's ability to tolerate faults and continue operating properly.

FDS: Functional Design Specification

FT: Flow Transmitter

GDP: Good Documentation Practices

GAMP: Good Automation Manufacturing Practice

GEF: Good Engineering Fact

GEP - Good Engineering Practice: is where established engineering methods and standards are applied throughout a project's 'life cycle' to deliver appropriate, cost effective solutions.

Generic Training: information learned from one system is brought to another similar type system.

GMP: Good Manufacturing Practice's are guidelines that outline the aspects of production and testing that can impact the quality of a product.

Hazop: Hazard and operability study: is a methodology for identifying and dealing with potential problems in industrial processes, particularly those which would create a hazardous situation or a severe impairment of the process.

HMI - Human Machine Interface: the user interface is (a place) where interaction between humans and machines occurs. The goal of interaction between a human and a machine at the user interface is effective operation and control of a machine and feedback from the machine itself which aids personnel in making operational decisions.

HP: Horse Power

HR: Human Relations/Resources

Humidity: is the amount of moisture in the air.

HVAC: Heating, Ventilating & Air Conditioning

i.e.: is a Latin expression meaning "that is" which written out fully in Latin is 'id est'.

In situ: is a Latin expression meaning "in the place" e.g. In the aerospace industry, equipment on board an aircraft must be tested in situ, or in place, to confirm everything functions properly as a system.

Interlock: safety/mechanical interlocks are put in place to prevent circumstances, which pose a risk to personnel, the environment and equipment. Electrical interlocks provide electrical isolation by means of auxiliary contacts in relays and magnetic motor starters. Process interlocks are used to prevent things happening that would affect the quality of the product.

Instrument: A device for recording, measuring or controlling a physical characteristic of a system, e.g. temperature, pressure, flow rate, etc.

IQ - Installation Qualification: is the documented verification that a system has been installed as per the approved design, manufacturer's instructions and the systems owner's requirements. Equipment components are identified and checked against the manufacturers' component listing. The working environment conditions are documented and checked to ensure that the components are suitable for the operation of the equipment itself.

ISO: Isometric

Isolating Device: an isolating device is always a mechanical unit that physically blocks or interrupts the flow or release of hazardous energy e.g. Electrical – a circuit breaker that interrupts all phases, Mechanical – locking bolt or brake, Hydraulic – line valve or spade blank.

Isolock: an isolock is a device that enables several people to lock out the isolation device with their own lock.

IT: Information Technology

Knowledge: facts or experiences known by a person, a state of knowing specific information on a subject.

KPI: Key Performance Indicator

KVA: Kilo - Volt - Ampere

KWh: Kilo - Watt - hour

LOPA - Level of Protection Analysis: is a methodology for hazard evaluation and risk assessment.

LT: Level Transmitter

LV: Low Voltage

Maintenance: is the in depth inspection of a plant's machinery and associated components and assemblies.

MCB: Miniature Circuit Breaker

MEWP: Mobile elevated working platform

MSDS - Material Safety Data Sheet: contains basic information intended to help personnel work safely with a material.

MTBF: Mean Time Between Failures.

MTTR: Mean Time To Repair.

MV: Medium Voltage.

N.B.: is a Latin expression that means 'Note Well' which written out fully in Latin is 'Nota Bene'.

Non-Critical Instrument: is an instrument where failure or inaccuracy will not result in an EH&S or GMP non-compliance. Instruments are typically operational instruments used for commissioning or engineering purposes.

OEE - Overall Equipment Effectiveness: is a critical methodology to drive improved efficiency, equipment availability, performance, higher quality and reduced costs for companies who are looking to maximise productivity while minimising operational costs. It is a powerful KPI in providing a metric that can be used by operations, production, engineering, maintenance, quality and continuous improvement teams.

OEM: Original Equipment Manufacturer.

Operational Efficiency: represents the life-cycle cost-effective mix of preventive, predictive and reliability-centred maintenance technologies, coupled with equipment calibration, tracking, and computerised maintenance management capabilities all targeting reliability, safety, occupant comfort, and system efficiency.

OPEX: Operational Expenditure.

Opinions: are preferences, beliefs and points of view.

O&M: Operation and Maintenance.

OQ - Operation Qualification: is the documented verification that the installed equipment or systems will operate throughout the design range. Equipments functions are checked to ensure that they conform to the manufacturer's specifications. This includes the use of certified, traceable simulators and standards to verify that the equipment and its process instruments are processing input signals correctly.

OSH: Occupational Safety & Health.

ODS: Ozone Depleting Substance.

PAT - Portable Appliance Testing: is to ensure that portable and transportable equipment are maintained in a safe condition so as to avoid any hazard to person's or property. It is carried out once a year on every portable appliance i.e. any appliance that has a plug fitted to it into a mains socket e.g. Laboratory equipment, power tools, office equipment. The results must be held for at least 5 years and must be made available for inspection by inspectors of the Health and Safety Authority.

PDA: Personal Digital Assistant i.e. digital note taking

Permit Issuer: Person responsible for the issuing and control of the permit.

Permit Holder: Person completing the work under the conditions of the permit.

Person in Charge: An occupier of an installation or premises or a manager, supervisor or operator of an activity or a suitably qualified and experienced deputy who is present on the installation at all times during its operation.

PFD: Process Flow Diagram

P&ID: Process/Piping and Instrument Drawings

P.I.D. - Proportional. Integral. Derivative: The PID terms of a PID controller are just numbers associated with a particular loop and they are adjusted by engineers during commissioning to give optimal loop performance which in turn gives optimal process performance, e.g. a PID loop could be used to control liquid level in a vessel or its associated jacket temperature, pressure or agitator speed to a set point set in a controller in a very short space of time with very little oscillations in the process it is controlling i.e. a set point of 80°C is set in the controller, temperature control is found to be erratic, fluctuating between 75- 85°C, 'see-saw effect' is seen on the PC trends, a properly tuned PID loop will remove this 'see- saw effect' and give an 80°C set point very quickly and accurately.

PLC: Programmable Logic Controller

PM: Project Management

PM - Preventative Maintenance: e.g. a check on the functionality of an instrument to confirm its functionality.

Power - Symbol (P) - measured in Watts - Electrical formulae: $(V * I)$ or $(R * I^2)$ or (V^2 / R)

PPE: Personal Protective Equipment

Problem solving: is the ability to identify and define problems as well as to generate effective solutions.

Proportionate Valve: (also known as a control valve), which is driven by an analogue output from a control system, could be made to open 20%, or by any other amount from 0 to 100% of fully open.

PQ - Performance Qualification: provides documented evidence that the equipment and ancillary systems when linked together can perform effectively and reproducibly to satisfy the requirements of the intended engineering/manufacturing process.

PRV: Pressure Relief Valve

PRV: Pressure Regulating Valve

PT: Pressure Transmitter

PTC: Positive Temperature Coefficient

PTW : Permit to Work

PWO: Plant Wide Optimisation

QA: Quality Assurance

RCA - Root Cause Analysis: is a structured process to study and learn from failures that do occur. Problems are best solved by attempting to correct or eliminate root causes, as opposed to merely addressing the immediately obvious symptoms. By directing corrective measures at root causes, it is hoped that the likelihood of problem reoccurrence will be minimised.

RCD: Residual Current Device.

Redundancy: means duplication or triplication of equipment that's needed to operate without disruption if primary equipment fails during operation.

Resistance symbol (Ω) - measured in Ohms - Electrical formulae: **(V / I)** or **(V^2 / P) or (P / I^2)**

Resistance decade box: or resistor substitution box is a unit containing resistors of many values, with one or more mechanical switches which allow any one of various discrete resistances offered by the box to be dialed in to an instrument.

Resistance thermometers: also called resistance temperature detectors or resistive thermal devices (RTD's) are temperature sensors that exploit the predictable change in electrical resistance of some materials with changing temperature.

RONA: Return On Net Assets.

RSJ: Rolled Steel Joist

SAT: Site Acceptance Test

SCADA: Supervisory Control And Data Acquisition

SDS: Software Design Specification

SEOR: Safety & Environmental Observation Report

SF: Service factor - is a measure of periodically overload capacity at which a motor can operate without overload or damage.

Service Provider / Consultant: are usually a firm or individual that provides a service either on or off-site and are employed to provide a service, professional advice and/or training based on their professional experience and academic qualifications.

Shadow board: contains outlines of designated tools to show where they should be stored.

SIL - Safety Integrity Level: is defined as a relative level of risk-reduction provided by a safety function, or to specify a target level of risk reduction.

SIS - Safety Instrumented system: performs specified functions to achieve or maintain a safe state of the process when unacceptable or dangerous process conditions are detected. Safety instrumented systems are separate and independent from regular control systems but are composed of similar elements, including sensors, logic solvers, actuators and support systems.

SOP: Standard Operating Procedure

Sparge Pipe: A horizontal pipe having fine holes drilled throughout its length so as to deliver a spray of water.

SRM: Standard Reference Method

Steam: Steam is efficient, economic to generate and provides excellent heat transfer. It is one of the most widely used media to convey heat over distances. It can hold five or six times as much potential energy as an equivalent mass of water. When water is heated in a boiler, it begins to absorb energy, depending on the pressure in the boiler; the water will evaporate at a certain temperature to form steam. Steam contains a large quantity of stored energy which will eventually be transferred to the process or the space to be heated. It can be generated at high pressures to give high steam temperatures i.e. the higher the pressure, the higher the temperature. More heat energy is contained within high temperature steam so its potential to do work is greater.

STP: Standard condition for temperature and pressure.

State Based Control: is a method for designing plant automation based on the principle that all process facilities operate in a recognised, definable process state. These cover normal and abnormal conditions of the process. State Based Control provides an environment for knowledge capture directly into the control design.

System Owners: The system owners are the equipment owners responsible for the daily operations and use. They are also responsible for ensuring that there is appropriate participation during the CAT assessments and for approving the critical instrument list upon completion of the assessments.

Template: a stencil, pattern or overlay used in graphic arts (drawing, painting, etc) and sewing to replicate letters, shapes or designs.

Thermocouple: is a junction between two dissimilar metals that produces a minute voltage in proportion to its temperature. Thermocouples are a widely used type of temperature sensor for measurement and control.

TI: Thermal Imaging

Torque: is a measure of how much a force acting on an object causes that object to rotate. It is a turning force applied to a shaft, tending to cause rotation. Torque is normally measured in pound/feet and is equal to the force applied times the radius through which it acts.

TT: Temperature Transmitter

UPS - Uninterrupted Power Supply: it's vital to provide continuous power. Bad power can cause unexpected behaviour to running microprocessor- based equipment. Therefore, the control system is only as reliable as the power provided to it. The key is to attach the output power of the UPS to the primary controller, which filters surges and minimises system recovery when power is re-established.

URS - User Requirement Specification: is raised by the system owner at the very beginning of the project stating what the system owner requires the equipment to do, the URS should express requirements and not design solutions.

Voltage - symbol (V) - measured in Volts - Electrical formulae: **(R * I)** or **(P / I)** or **√(P x R)**

VSD - Variable Speed Drive: is an electronic device that controls the characteristics of a motor's electrical supply allowing the speed and torque of the motor to be matched with the requirements of the machine it is driving. VSD's regulate the operation of electric motors and save energy by matching the output of motor-driven pumps, fans, conveyors and similar equipment with the actual demand of the systems they support. In addition to saving money, drives can also improve process control, whilst reducing waste, maintenance costs and carbon emissions.

WI: Work Instruction

Zeroth's Law states: If two thermodynamic systems are in thermal equilibrium with a third, they are also in thermal equilibrium with each other.

ZNE: Zero Net Energy - The ZNE consumption principle focuses on renewable energy harvesting as a means to cut greenhouse gas emissions. E.g. a building that contributes more than it consumes by installing energy generating technologies e.g. wind, solar.

20
SUMMARY

Preparation, if qualified personnel are coming to site to carry out a specialist role or task whether they are consultants, engineers, tradesmen etc. the company's key personnel must ensure to have it clear in their heads what they want done and what they want them to do. Be clear and concise in their instructions and leave no room for ambiguity. Have the preparation work done and tell them exactly and how it is to be done citing site safety procedures and also how they must leave the area exactly as they found it.

Ensure all potential obstacles, both physical and red tape are dealt with and agreed by all the relevant onsite affected parties prior to any proposed work being carried out. Having contract personnel standing around while trying to get the 'go ahead' to work on a piece of equipment because of lack of preparatory work on the company's behalf can be very expensive and possibly dangerous to the personnel carrying out the work due to the lack of an appropriate hazop or method statement of the intended work being considered and written.

Tip: Whether you have rectified a minor or major fault on a piece of equipment, replay your approach in your head on how you resolved the problem, write down the sequence, review and learn from the experience. Ask yourself, would you take a different approach the next time and maybe save valuable downtime and the use of costly parts.

N.B. With an incomplete understanding of a problem, it is very easy to jump to the wrong conclusions.

Understanding, Understanding, Understanding, cannot be stressed enough and is vital in maintaining good trouble shooting and fault finding techniques.

The better the access to knowledge, the better informed decisions can be made.

It is the difference between diagnosing and misdiagnosing a problem. Personnel have a much better chance of preventing any future potential problems or diagnosing an actual fault quicker, if they understand exactly how each component associated with the piece of equipment operates and how the individual components combine to make the machine do what it was designed to do. If possible, try to discover when a fault first started to manifest itself, use technology where possible especially if the pieces of equipments' parameters are monitored by a computerised trending system.

Great credit to all for resolving a fault and getting the plant back on line, more credit though for sourcing the cause of the problem and implementing the follow up actions to prevent a reoccurrence.

5 key learning points:

1. Keep focus on safety.
2. Minimise downtime.
3. Avoid risk.
4. Increase efficiency.
5. Save money.